Kirstenbosch Gard‹

GROW
DISAᴗ

A PRACTICAL GUIDE TO THE CULTIVATION
AND PROPAGATION OF EVERGREEN
AND DECIDUOUS *DISA* SPECIES OF
SOUTHERN AFRICA

Text by Hildegard Crous and Graham Duncan

Photographs by Hildegard Crous, Graham Duncan
and from the Kirstenbosch Collection

Disa racemosa flowering *en masse* after a fire in the southern Cape Peninsula

Dedicated to the memory of
Dr Louis Vogelpoel
1922 – 2005

Dave van der Merwe

In recognition of his tremendous contribution to the propagation and cultivation of disas, which he made with great enthusiasm and dedication.

Louis Vogelpoel had an exceptional career as an outstanding general physician, clinical cardiologist and horticultural scientist. After graduating at the University of Cape Town (UCT), with first class honours in 1945, he completed an internship at Groote Schuur Hospital and was awarded fellowships which allowed him to spend two years at the National Heart Hospital in London. He made a major contribution to the Department of Medicine and the Cardiac Clinic at UCT, and as a gifted and enthusiastic teacher, he was instrumental in the training of generations of undergraduate medical students.

Louis had another life outside medicine. He was recognized as a horticultural scholar and researcher. His interest was mainly in indigenous South African flora and he was regarded as an expert on ericas and South African orchids, especially the genus *Disa*. After his first publication on the propagation and cultivation of *Disa uniflora* in *The American Orchid Society Bulletin* in 1980, many more papers followed on various aspects of disas, especially their breeding and analysis of their flower pigments. He was also an expert photographer of disas. His superb garden at his home reflected his interest, hard work and expertise, and he was generous in sharing his plants with friends. Many of us are fortunate to have plants that he gave us growing in our gardens. What a wonderful way to be remembered!

Prof. Pat Commerford, Head of Cardiology, University of Cape Town.
In Memoriam, South African Medical Journal, *November 2005*

A superb mixed colony of *Disa maculata* (blue) and *D. virginalis* (white) flowering on a mossy bank near the Hex River Mountains in early summer (see page 69)

Percy Sargeant

CONTENTS

Below: The pink form of *Disa uniflora* in full flower on a mossy bank in the Hex River Mountains (see page 44)

Opposite: The hybrid *Disa* Veitchii (see pages 4, 54, 55)

John Winter

A BRIEF HISTORY

"...its gorgeous colouring and beauty...can be fully appreciated only when seeing the blooms on a cliff, just touched by the morning sun, or on the banks of a murmuring brook whose clear water reflects their glory."

Rudolph Marloth

"Here the great disas, hovering o'er the springs, Gaze with delight upon their mirrored wings."

Vine Hall

The genus *Disa* was established in 1767 by the Swedish physician and botanist Petrus Jonas Bergius (1730-1790) who later became Professor of Natural History and Pharmacy at the *Collegium Medicum* in Stockholm, Sweden. He provided no explanation for his choice of name, however one theory suggests that upon seeing the flower of *Disa uniflora*, the red disa, Bergius was reminded of the mythical Queen Disa of Sweden. When the King of the Sveas was looking for a Queen, he asked for ladies to be presented, but they were to be neither dressed nor naked. The beautiful Disa presented herself wrapped in a fishing net, and a similar net-veined pattern is clearly visible on the median sepal of *Disa uniflora*. An alternative explanation is that he named it from the Latin word *dis,* meaning rich or opulent, referring to the colouring of the flower that has a golden sparkle when seen in bright sunlight.

The startling beauty of this species makes it one of the most striking flowers in the Fynbos Biome, and it has for many years been referred to as the 'Pride of Table Mountain'. Prior to the advent of the modern binomial system of classification for plants and animals introduced by Carolus Linnaeus in his *Species Plantarum* in 1753, *Disa uniflora* had been described for the first time in 1704 by the English priest John Ray (1628-1705), using the phrase name *Orchis africana flore singulari herbaceo,* in the supplement to his *Historia plantarum generalis*. The species name '*uniflora*' is however somewhat misleading, as specimens frequently produce three or more flowers per inflorescence. Bergius apparently only ever saw specimens with single flowers.

Disa tripetaloides (see page 49)

Once very common along most streams on Table Mountain, today *D. uniflora* is confined to the more inaccessible wet ledges and rock faces. It was especially plentiful along cold stream banks of the southern and eastern slopes, but unfortunately the construction of the Woodhead Reservoir led to the destruction of large numbers of these plants. Sir Percy Fitzpatrick, South African naturalist, author and politician, once commented that the plants covered the area of this reservoir 'like a scarlet blanket'. *D. uniflora* has the largest flower of all the South African orchids and is considered by many as the most beautiful orchid in the world.

Since the late nineteenth century, *D. uniflora* has been extensively used in hybridization experiments; the first hybrid in which it was used became known as *Disa* Veitchii and was the result of a cross with another species from the Western and Eastern Cape, the deep pink-flowered *Disa racemosa*, raised in England in 1891 in the nurseries of James Veitch and Sons. Numerous subsequent hybrids using these two species and their progeny, as well as *Disa tripetaloides,* later appeared (see table on page 55).

Probably the most frequently seen evergreen *Disa* in the southern Cape is *D. tripetaloides*. The first scientific collection of the plant was made in 1772 by the Swedish botanist Carl Peter Thunberg, often referred to as the 'Father of South African Botany'. He probably found it in the Langkloof Mountains near Joubertina where this species is still commonly encountered. Its complicated taxonomic history began with it first being described as a species of *Orchis, O. tripetaloides,*

The Table Mountain form of *Disa uniflora*, reproduced from a watercolour painting by Fay Anderson, from specimens cultivated by Dr Louis Vogelpoel (see page 44)

in *Supplementum plantarum* in 1781 by Carolus Linnaeus junior, son of the famed Carolus Linnaeus. It underwent several further name changes until finally in 1879 it received its present name of *Disa tripetaloides*. Its relative ease of cultivation and propensity to hybridization resulted in it becoming firmly established in cultivation in Britain by the 1880s, and by 1900 it had been used in the creation of two primary hybrids, *Disa* Kewensis (a cross with *D. uniflora*) and *Disa* Langleyensis (a cross with *D. racemosa*). From these two hybrids, numerous secondary hybrids arose.

Another noteworthy species discovered during the eighteenth century was the white- or pale mauve-flowered *D. sagittalis*, found by the Swedish botanist Anders Sparrmann in the late 1770s in the southern Cape. It is a winter-growing plant and one of the most easily maintained disas in cultivation (see page 73).

A rare, bright yellow-flowered *Disa* first discovered by the English explorer William Burchell on the southern slopes of the Langeberg Mountains of the southern Cape in 1815, was later described as a variety of *D. tripetaloides* in 1893. It was subsequently upgraded to subspecies level in 1981, and finally in 1993 it became recognized as a species of its own: *Disa aurata* (see page 50).

Previously regarded as a red-flowered form of *Disa tripetaloides*, the robust and beautiful *D. cardinalis* is restricted to the inland slopes of the Langeberg Mountains near Riversdale and was formally described as a species as recently as 1980. It was introduced into cultivation in the late 1970s and the outstanding primary hybrid *D.* Kirstenbosch Pride (a cross with

Disa uniflora), developed by former Kirstenbosch Curator John Winter, was registered in 1981 (see pages 10, 50–51).

Amongst the startling array of deciduous, winter-growing disas of southern Africa, perhaps the most beautiful are the blue-flowered *Disa graminifolia* and *D. purpurascens* described in 1826 and 1884 respectively (previously known under the genera *Herschelia* and *Herschelianthe*). Established by John Lindley in 1838, the frequently blue-flowered genus *Herschelia* (the genus *Herschelianthe* was established by Rauschert in 1983, nearly 150 years later) was appropriately named for the nineteenth century British astronomer Sir John Herschel (1792-1871) because the flowers of *D. graminifolia* were coloured 'as intensely blue as the southern skies'. Herschel and his wife Lady Herschel spent a five-year period (1834-1838) at the Cape surveying the southern skies and depicting many of the floral treasures of the Cape in beautiful pencil and water-colour studies. Once common on Table Mountain, *D. graminifolia* is now only to be found occasionally on the upper slopes, probably due to indiscriminate picking in the past.

Amongst the summer-growing disas, one of the most noteworthy species collected in the early nineteenth centuries was the spectacular pink-flowered *D. crassicornis*, collected by the German explorer J.F. Drege in the Witteberg Mountains of the Eastern Cape. Unfortunately, long-term success with *D. crassicornis* in cultivation and most other summer-growers has yet to be achieved.

In 1895 the pharmacist, botanist and naturalist Dr Rudolph Marloth, often regarded as the founder of modern botany in South Africa, made his

Disa Kewensis, a hybrid between *D. uniflora* and *D. tripetaloides*, was first raised in Britain in 1900. The above specimen was cultivated at Kirstenbosch (see pages 6, 55)

startling discovery of the pollinator of *D. uniflora,* after capturing a butterfly on Table Mountain. This very large and colourful butterfly (*Aeropetes tulbaghia,* previously known as *Meneris tulbaghia*) that became known as the Table Mountain beauty, had pollen from the red *D. uniflora* stuck to its legs (see pages 44, 80, 81).

As a result of the pioneering work carried out at the Cape in the late twentieth century by two seasoned growers of evergreen disas, Mr Helmut Meyer of Stellenbosch and Mr K.C. Johnson of Somerset West, the keys to successful cultivation and propagation of evergreen species over the long term were brought to light. In 1962 Johnson 're-made' *Disa* Veitchii, and during the period 1981-1996, no fewer than 120 new hybrids were registered in South Africa. The inflorescences of evergreen *Disa* species and hybrids can last several weeks in the vase and they are also potentially good pot plants. Today evergreen disas, and their hybrids in particular, are being grown successfully in a number of countries outside southern Africa, notably in Australia, New Zealand, the United Kingdom and the United States of America. Although not easily maintained in areas with hot and humid summer climates, such as Japan, they are nevertheless being grown to perfection in that country's somewhat cooler region of Hokkaido Island.

Perhaps not surprisingly, *Disa uniflora* has served as the emblem of the Mountain Club of South Africa since its inception in 1891, and is also the emblem of the Western Province Rugby Football Union and the floral emblem of the Western Cape.

Percy Sargeant

Top: *Disa* Veitchii (see pages 4, 55)

Below: *Disa graminifolia* (see page 68)

Above: The hybrid *Disa* Helmut Meyer (see pages 55, 56)

Below: *Disa cardinalis* flowering along a perennial stream in the Langeberg Mountains in the southern Cape (see page 50)

Opposite: *Disa uniflora* (see page 44)

Rodney Saunders

GENERAL INFORMATION
Taxonomy

The family Orchidaceae is the largest of the monocotyledons worldwide and is handsomely represented in southern Africa. We have the horticulturally important *Disa, Ansellia* and *Eulophia* included among 53 genera and approximately 470 species, most of which are endemic. Most of the orchids found here are terrestrial and either evergreen or deciduous, being either winter- or summer-growing. There are comparatively few epiphytic species occurring in the moister areas. The greatest diversity in orchid species is found in the Western Cape, an area with a Mediterranean-type climate, where mild, wet winters alternate with long, hot, dry summers and flowering takes place mainly in spring and early summer. By far the largest group in this region is *Disa*, with about 93 species including *Disa uniflora*, the red disa, a plant of truly classic beauty that has without doubt made the greatest contribution to horticulture of all the southern African orchids. In total, *Disa* comprises in excess of 160 species. The genus is distributed throughout the fynbos and grassland areas of South Africa, with some species occurring in the drier inland areas, and it reaches further north into tropical Africa, with one species occurring on the Island of Reunion and another occurring on the Arabian Peninsula.

11

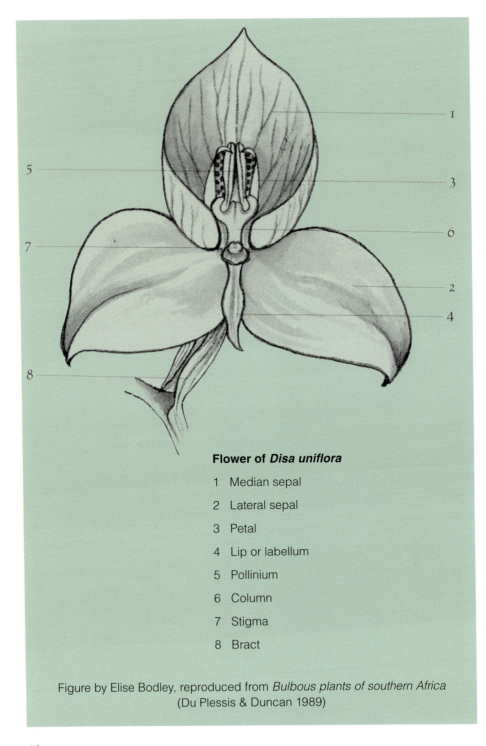

Flower of *Disa uniflora*

1 Median sepal

2 Lateral sepal

3 Petal

4 Lip or labellum

5 Pollinium

6 Column

7 Stigma

8 Bract

Figure by Elise Bodley, reproduced from *Bulbous plants of southern Africa*
(Du Plessis & Duncan 1989)

Together with the small genus *Schizodium,* which occurs entirely in the winter-rainfall area of southern Africa, *Disa* belongs to the subtribe Disinae which contains about 170 species. *Disa* was taxonomically revised by Prof. Peter Linder in 1981 in which he retained *Herschelianthe* and *Monadenia* as separate genera within the subtribe. However recent research based on anatomical, morphological, palynological and molecular data has indicated that *Herschelianthe* and *Monadenia* are more appropriately placed in *Disa*. This new classification was presented in the beautifully illustrated, standard reference work *Orchids of Southern Africa*, co-authored by Prof. Peter Linder and Dr Hubert Kurzweil, in 1999.

Structure of the Disa flower

The structure of the *Disa* flower is best illustrated in *Disa uniflora*, as it has the largest flower of all the southern African orchids (see figure opposite). The flowering stem consists of the peduncle, which supports the stalk of the inflorescence, and the rachis, an extension of the peduncle that bears the individual flowers. As with all orchids, the flowers of disas are zygomorphic or irregular, meaning they can be bisected into identical halves in one plane only.

Flower colour in disas is extremely wide-ranging, from pure white (*D. vasselotii*) through innumerable shades of red and pink to blue, violet and even maroon-black (*D. lugens* var. *nigrescens*). However the most colourful parts of the *Disa* flower are not the petals as in many other orchids, but the sepals. The petals in *Disa* are greatly reduced and appear alongside the pollinia. On the outer perimeter of the flower, three sepals occur which are often brightly coloured. The median sepal is situated at the top of the flower, and two sepals are arranged on the sides of the flower, known as lateral sepals. The median sepal is sometimes elongated to form a short or long, tubular spur. Two petals occur at the sides of the flower; they are basally fused with the column and sometimes obscured inside the median sepal, while another segment known as the lip or labellum occurs at the base of the flower, usually in the middle.

The lip of the *Disa* flower is very variable and may be simple as in *D. uniflora*, or deeply dissected and beard-like as in *D. lugens* (see page 28). In the centre of the flower, a single stamen is fused together with the stigma and style to form a rigid organ called the column. Within the anther, the pollen grains are stuck together to form large masses known as pollinia, which occur in pairs and are bright yellow and attached to sticky glands. The stigma occurs as a rounded sticky surface at the base of the column. The flower is borne on a pedicel that is enclosed by a leaf-like bract (see figure opposite).

Anthony Hitchcock

A pink form of *Disa uniflora* flowering in the Hex River Mountains (see page 44)

Left : *Disa marlothii* is a rare, evergreen species from the inland mountains of the Western Cape and the extreme south-western part of the Eastern Cape

Below left: *Disa uniflora* occurs in the south-western part of the Western Cape and is an evergreen species found in perennially wet seepages over cliffs and along perennial streams (see page 44)

Below right: *Disa tripetaloides* is a widespread, evergreen species occurring along perennial streams and moist mountain slopes from the southern part of the Western Cape to the southern part of KwaZulu-Natal (see page 49)

Growth cycle

There are three basic growth cycles in *Disa* species: the evergreen cycle in which plants occur in seepage and streamside conditions, the winter-growing cycle in which the plants are deciduous, growing during winter and undergoing a dormant period over summer, and the summer-growing cycle in which the plants grow during the summer and undergo a dormant period over the winter months.

Evergreen disas

Evergreen disas occur mostly in the Western Cape at altitudes varying from close to sea level to approximately 1 200 m above sea level. These species, *D. uniflora, D. tripetaloides, D. cardinalis, D. aurata, D. caulescens, D. marlothii* and *D. uncinata*, usually colonize perennial mountain streams alongside deep pools and indentations where the water flow is not too rapid. Other preferred habitats are on wet, shady, south-facing rock faces and under the drip-line of waterfalls amongst moss and roots, clinging precariously to rock ledges. As the streams are mostly fed by run-off water, the water quality is directly influenced by the vegetation type it runs through, namely the fynbos. This vegetation type is unique to the winter rainfall region and renders the water acid, also giving it its distinctive brown, tea-like colouration. A pH value as low as 3.55 is not uncommon (see page 41). Another factor affecting the pH value of the water is the Table Mountain Sandstone rock substrate that is dominant in the Fynbos Biome. This substrate also has an acidic pH value and is very poor in nutrients. Disas grow in sand derived from Table Mountain Sandstone with organic silt trapped in it, or in organic fibrous humus that is well impregnated with sand. Both substrates have extremely good drainage properties, as well as the required low pH value and low nutrient content.

Unlike most other disas, the plants are never truly dormant. During the cold winter rainy season the growth rate slows down and plants are often submerged under water while the rivers overflow. Disas survive such waterlogged conditions without rotting only because the water temperature plummets to as low as 4 °C and sometimes even lower. Vegetative growth begins at the end of the flowering season in late summer and early autumn with the formation by the mature plant of a new storage organ known as a tuberoid (that lasts about one year) next to the old tuberoid that gradually withers and dies. In addition, one or more stolons may be produced from the base of the plant that give rise to new plants a distance away from the mature plant (see page 89). New plantlets may also develop around the immediate base of the mature plant. Throughout the cold winter months vegetative growth continues slowly, increasing as temperatures rise from early spring onwards, culminating in the development of flowers from early summer to midsummer (October to March in the southern hemisphere). Although many streams in the Western Cape are perennial, the stream topography, water quality and climatic conditions also need to be appropriate for disas in order for the plants to grow successfully. Providing these conditions artificially is reasonably easy, hence the successful cultivation of these disas in the past.

Opposite left: *Disa tripetaloides* (see page 49)

Opposite right: *Disa uniflora*, an evergreen species (Cederberg form, see page 44)

Opposite below: *Disa uniflora*, (Franschhoek form, see page 44)

Below: *Disa uncinata*, an evergreen species commonly encountered along perennial mountain streams in the south-western and southern parts of the Western Cape (see page 15)

Peter Jackson

17

Disa cornuta is a deciduous, winter-growing orchid with a wide distribution in southern Africa extending from the Western Cape to Zimbabwe (see page 65)

18

Although the above-mentioned species do not occur together naturally, their preferred habitats are similar and vary mainly in altitude and availability of water. The streams may shrink considerably during the dry summer months, but the plants have evolved in various ways to overcome these periods of drought. *Disa uniflora* has developed a large storage tuberoid, whereas *D. tripetaloides*, *D. cardinalis* and *D. aurata*, in addition to having tuberoids, have developed thick fleshy roots which augment the function of storage tuberoids.

Disas are cool-growing orchids and thus need a cool, well-ventilated environment. The mountainous areas of the Western Cape provide exactly these conditions. During winter, temperatures drop to below freezing for short periods of time (see page 36). The mountains are exposed to the full force of north-westerly winds that bring the winter rain. Summers are hot and dry with temperatures reaching well above 30 °C, although summer temperatures can also fall as low as 8 °C. As long as the streams keep flowing, providing a cool environment for the roots, the plants are able to survive. A phenomenon that dramatically affects the supply of water and the survival of disas during these hot, dry months is the 'tablecloth' often seen draped over the Western Cape mountains. During summer, moisture-laden winds blowing northwards from the Antarctic are forced to rise and cold, moisture-laden clouds settle on the higher mountain ranges, forming the spectacular 'tablecloth' effect. As a result, moisture is deposited and drenches the mountain tops. Temperatures in the clouded areas fall, and freezing conditions have been known to occur at night in the mountains in midsummer.

Winter-growing, summer-dormant disas

The leaves of most winter-growing disas and other deciduous terrestrials such as *Bartholina*, *Holothrix* and *Pterygodium* from the winter rainfall areas of South Africa are synanthous (flowers and leaves appear together) and begin growth in early autumn as night temperatures drop after the long, hot, dry summer. Well known species include *D. cornuta*, *D. draconis*, *D. ferruginea*, *D. graminifolia*, *D. racemosa*, *D. sagittalis*, *D. spathulata* and *D. venosa*. Firstly, new leaf shoots develop from the dormant tuberoid, followed by new adventitious roots produced at the base of the leaf shoots that grow throughout the winter months. In most species, flower shoots develop from mid spring to early summer, followed by a pronounced dormant period as temperatures rise sharply in early summer. The old tuberoid is replaced every year and the new replacement tuberoid develops during the winter growing period. *D. racemosa* is known to produce multiple tuberoids, thereby forming dense populations of plants when the growth season commences. *D. cornuta* and *D. sagittalis* are unusual in that their distribution extends over two distinct geographical areas; winter rainfall and summer rainfall areas, yet in both of these they follow a mainly winter-growing cycle. *D. cornuta* starts its growth cycle in autumn, flowers from spring to midsummer and goes dormant in late summer, while *D. sagittalis* commences growth in midwinter, flowers from spring to early summer and goes dormant in summer.

Opposite: *Disa nervosa* is a deciduous, summer-growing species occurring in dry, stony grassland from the Eastern Cape to Mpumalanga (see page 22)

Right: *Disa polygonoides* occurs in grassland or seasonal marshes from the Eastern Cape to Limpopo and is a deciduous, summer-growing orchid (see page 22)

Below right: *Disa amoena* is a deciduous, summer-growing orchid restricted to dry, stony grassland mountain slopes in northern Mpumalanga

Cameron McMaster

21

Summer-growing, winter-dormant disas

The growth cycle of the deciduous summer-growers progresses in much the same way as the winter-growers, except that growth starts in spring, and dormancy sets in during autumn and early winter. Some of the better known species include *D. crassicornis, D. nervosa, D. polygonoides, D. porrecta* (see photo below), *D. saxicola* and *D. woodii*. At the beginning of the growing season a shoot develops which forms the regenerative bud. As the shoot develops, adventitious roots grow from the base of the stem where it is attached to the tuberoid. A replacement tuberoid develops during the growing period but in certain summer-growing species occurring in grasslands that are regularly burned, the tuberoid may reach an age of two years or more. Although most summer-growing disas are synanthous, certain species such as *D. baurii* have hysteranthous leaves that grow shortly after the flowers have appeared.

Below: The deciduous, summer-growing orchid, *Disa porrecta* flowers in early autumn and occurs in grassland in the Eastern Cape, Free State and Lesotho

Opposite: A field of *Disa racemosa* flowering after a fire on the southern Cape Peninsula (see page 72)

Cameron McMaster

Distribution and habitat

Southern African disas can be broadly divided according to the geographical areas where they mostly occur, that is, winter rainfall and summer rainfall disas. The winter rainfall species can be further categorized into two groups: evergreen and deciduous (winter-growing, summer dormant) species. Within southern Africa, disas occur in all nine provinces of South Africa as well as in Lesotho and Swaziland, but are absent from Botswana and Namibia.

The distribution of winter rainfall disas extends from the north-western part of the Northern Cape in the Kamiesberg, southwards throughout the Western Cape to the south-eastern part of the Eastern Cape around Port Elizabeth. Most of the species occurring in the Western Cape are endemic to that province, which experiences a mainly Mediterranean climate where rains fall during the cool winter months and summers are long, hot and dry, except in mountainous areas where mist and low southeaster cloud provide cool conditions for most of the year. The evergreen species are found mainly in mountainous terrain along perennial streams and in seepages, swamps or seasonally wet habitats. The most widely distributed evergreen disa is *D. tripetaloides*, which occurs from the southern parts of the Western Cape to the southern parts of KwaZulu-Natal, from sea level up to 1000 m (see page 49).

The deciduous, winter-growing disas occur in a wide range of habitats, from dry to marshy areas on sandy flats, to dry or moist rocky mountain slopes. One of the most widespread members of the winter-growing group is the striking

Percy Sargeant

golden orchid, *D. cornuta,* which occurs in well-drained areas in sandstone soils of the south-western parts of the Western Cape, to grassland in heavy soils in the northern parts of KwaZulu-Natal and further on to eastern Zimbabwe. It is one of very few winter-growing disas found in both winter and summer rainfall areas (see page 65). In the south-eastern parts of the Western Cape coast around Knysna, rainfall occurs mainly in winter although some summer rain falls, and here too several winter-growing *Disa* species are encountered including the widespread *D. cornuta* and *D. uncinata.*

The summer rainfall disas occur across southern Africa, stretching from the south-eastern parts of the Eastern Cape near Port Elizabeth in an easterly direction to Limpopo in the far north. Here rains fall predominantly during the hot summer period and the winters are cool to very cold and dry. The disas occur in a wide variety of habitats including forest margins and dry to moist grassland slopes, on rocky ledges or damp flood plains, and from sea level to alpine conditions at 3 000 m along the Drakensberg mountain range.

A remarkable feature of the distribution of southern African orchids in general, and disas in particular, is the exceptionally high level of endemism. Most of the species are not found outside their region, and due to great variation in terrain and microclimatic conditions, many *Disa* species have highly restricted ranges, such as *D. barbata,* which is confined to a very small area of the south-western Cape. Aspects favoured by southern African disas vary greatly from full sun (e.g. *D. spathulata* and *D. woodii*) to partial shade (e.g. *D. tripetaloides* and *D. cardinalis*), and

Theodore Würts

Wouter van Warmelo

even deep shade (e.g. *D. longicornu* and *D. maculata*). The genus is encountered mainly in acidic soils, and some deciduous species occur in widely differing soil types, such as *D. spathulata* that is found in both well-drained sandstone and moist shale conditions.

In a number of deciduous and evergreen disas of the Western Cape, flowering is greatly stimulated by periodic wild fires that sweep through thick fynbos vegetation in summer and autumn. The frequency of these fires is extremely erratic and may take place annually, every few years, or sometimes after an interval of up to 40 years. It would appear that smoke derived from burning plant material provides a chemical cue that may stimulate mass flowering in certain disas like *D. racemosa* and numerous other geophytes of this region, (such as *Cyrtanthus ventricosus*) as has been shown in the germination of seed of many fynbos species. In 2004, the major constituent of this chemical cue was identified by Australian and South African researchers as a butenolide compound.

Opposite: A clump of *Disa cardinalis* clings to a sandstone boulder amid a raging torrent in the Langeberg Mountains of the southern Cape (see page 50)

Above left: *Disa hians* flowers in midsummer and is commonly encountered in the south-eastern part of the Western Cape and the southern part of the Eastern Cape, in well-drained, gravelly soils

Left: *Disa harveiana* subsp. *harveiana* flowering in acid sandstone on Table Mountain

26

Opposite: *Disa venosa* flowering on a rocky sandstone slope (see page 75)

Left: *Disa tenuifolia* flowering in profusion after a fire on a damp sandstone slope near Elgin (see page 75)

Below: *Disa ferruginea* flowering between sandstone boulders on Table Mountain in midsummer (see page 67)

Conservation

There are several threats to the well-being of *Disa* and other orchid species in southern Africa today, the greatest of which is the destruction of their habitat; most notably for housing and industrial 'development' and road construction, especially in coastal areas. The spread of alien vegetation in the sensitive ecosystems in which many disas occur places an additional burden on the survival of many species. A number of species are endangered in their natural habitats, most of which are endemics of the winter rainfall areas of the Western and Eastern Cape. Perhaps the most critically endangered of these is *D. barbata*, a beautiful white to pale blue-flowered species from seasonally moist sandy flats in a highly restricted area of the south-western Cape. Similarly, numbers of the bizarre greenish *D. lugens* var. *lugens* have dwindled over much of its range in coastal parts of the Western and Eastern Cape due mainly to habitat destruction, while the near black-flowered *D. lugens* var. *nigrescens* has always been confined to a single population near Humansdorp in the Eastern Cape. Once very common on Table Mountain, numbers of *Disa uniflora* have steadily declined over the past century owing mainly to the construction of a reservoir and illegal picking. It is clear that a growing number of disas are in need of conservation.

Although the cultivation of many deciduous *Disa* species is, as yet, beyond the scope of the average gardener, and in many cases the skilled professional as well, every effort should be made to cultivate those species for which protocols are available and in this way make a contribution towards their conservation. All indigenous *Disa* species of southern Africa are protected by law and may not be removed from the wild without obtaining the necessary collecting permits from the relevant provincial Nature Conservation authorities. Those wishing to begin their own *Disa* collections are encouraged to obtain propagative material from registered orchid nurseries, by contacting the Disa Orchid Society of South Africa (see page 116).

Cameron McMaster

Opposite: *Disa barbata* is probably the most endangered orchid in South Africa (see pages 94–95)

Left: Population numbers of *Disa lugens* var. *lugens* have declined drastically due to the destruction of its coastal habitat (see page 71)

William Liltved

Below: A fine stand of the yellow form of *Disa uniflora* in the Kirstenbosch nursery (see pages 48, 88)

Opposite: *Disa uniflora* (see page 44)

CULTIVATION

It has to be said at the outset that success with the cultivation of disas requires dedication and constant attention to the requirements of the plants. Although many *Disa* species are difficult or even impossible to grow over the long term, here we provide cultivation notes on a selection of evergreen and deciduous species, most of which can be grown quite easily at home. The cultivation notes provided here are for southern hemisphere conditions; in order to obtain the corresponding month for northern hemisphere conditions, six months should be added.

To begin with, it is most important to clarify the plants' growth cycle. Are they evergreen or winter- or summer-growing (deciduous) species? To grow these orchids successfully means being able to keep them going over an extended period, ideally from seed to flowering stage. Although both evergreen and deciduous adult *Disa* plants need a continued mycorrhizal relationship in order to survive, the evergreen species appear to be relatively independent of a specific mycorrhizal fungus, while the deciduous species are very dependent on specific mycorrhizal fungi. Different species have different requirements, for example, some go dormant earlier than others and therefore an appropriate watering procedure is an essential requirement for successful cultivation. Disas are not tolerant of severe frost, and in northern hemisphere countries with cold winter climates, both evergreen and deciduous species require the protection of the cool greenhouse. The cultivation notes provided here serve only as a guide; each grower will discover his or her own ideal methods that suit local conditions.

Evergreen disas

Although evergreen disas often grow in full sun for much of the day in their natural habitat, in most nursery situations these conditions would be too severe. The golden rule for cultivating these species successfully is to keep the roots cool and moist. This is achieved by maintaining the heat build-up in the greenhouse at a suitably low level, optimally no higher than 24 °C, and if necessary, by providing additional shading.

Greenhouses for disas should be constructed to allow the accumulation of hot air inside the greenhouse to escape through the top. In the southern hemisphere where daytime summer temperatures are high, a minimum roof height of 4 m for a greenhouse is recommended. Structures can be enclosed simply with shade cloth or insect netting on the side walls, clear weatherproof material as a roof, and adequate shading in the form of 50% black shade cloth. The shade cloth should ideally be suspended above the roof with at least a 20 cm clearance between the roof and the shade cloth (it is however often more practical to erect the shading inside the greenhouse, the limitation being that more effort needs

to be made to reduce the increased heat build-up inside the greenhouse). This type of structure offers excellent ventilation and keeps rain off the plants. Should it be necessary to enclose the greenhouse all around with plastic or any other suitably weatherproof material, provide adequate ventilation with an extractor fan and wet wall system. Disas can also be grown in drip trays mounted against a wall (see photo opposite). Additional cooling can also be provided by installing a fogging system that activates at a pre-set temperature, e.g. 24 °C. The purpose of the fogging system is not to provide moisture for the plants, but to cool down the air temperature. Should this measure still not provide sufficient cooling, an extra layer of shade cloth should be added.

Correct watering procedure is as important as the growing environment (see detailed notes on watering procedure on page 40).

Potting methods

An ideal medium is a very well draining one, which can nevertheless retain moisture successfully yet not remain soggy. Soggy conditions can very easily become anaerobic, which always leads to root rot very rapidly. Various potting components and combinations thereof can be used with varying degrees of success. Coarse, inert materials like perlite, vermiculite, silica sand, river sand and polystyrene crumbs can all be used to 'lighten' the mix and enable better aeration and drainage. Absorbent materials like peatmoss, fine milled bark, fern fibre or coconut fibre are suitable for retaining moisture. Availability will to a large degree influence one's choice of material, as will the climate in which

Top: Evergreen disas can be successfully grown by the erection of metal brackets to hold drip trays against a wall, covered with shade cloth

Above: A 'planter tube' has proved a most successful method of growing winter-growing, summer-dormant disas like *Disa maculata* and evergreen disas (see page 71)

the plants are grown, the degree of air humidity, and average daytime temperatures. Whatever the situation and one's choice of growing medium, ensure that the medium has excellent drainage yet will be able to retain moisture, that the roots are kept cool, and that the pH value of the medium errs on the acidic side. A pH value between 5 and 6.5 is acceptable. Peatmoss, fern fibre and milled bark are all acidic (varying between 3.5 and 4.5) and can be most advantageous in keeping the pH value of a medium acidic if other influences, such as the water used, are not quite acidic enough.

For all potting mixtures, except where the preferred potting medium is sphagnum moss, the guideline of 70% coarse inert material and 30% absorbent material will provide the necessary texture.

The sphagnum moss method

Potting mixture: 2 parts sphagnum moss (well soaked, with excess water squeezed out) and 1 part coarse quartzitic gravel or polystyrene crumbs of 2-3 mm diameter or suitably graded perlite (this is added to the medium to prevent the moss from collapsing, the ratio of which can be increased according to necessity).

To pot up: Soak the dry sphagnum moss to saturation point. Squeeze out excess water and mix well with inert medium. Make sure the mixture is thoroughly mixed. Gently pack moss underneath and around the plant. Do not push or squeeze the moss around the plant. Place the pot in a deep saucer and fill with water to approximately 5 cm. Once the water has been soaked up naturally, allow the excess to drain.

This medium is ideal for areas where high daytime temperatures are not uncommon

Above: Sphagnum moss (left) and quartz gravel (right)

Disa Kirstenbosch Pride 'Clare' (see page 55)

Louis Vogelpoel

Dave van der Merwe

Some evergreen disas can withstand temperatures below freezing for short periods of time; here Brenda Anderson observes plants of the pink form of *Disa uniflora* hidden behind icicles in the Hex River Mountains (see page 19)

Opposite: Re-potting a *Disa* clump into quartz gravel potting mixture

(25-35 °C). As the water evaporates off the surface of the moss it has a cooling effect, keeping the roots cool.

Quartz gravel method

Potting mixture: 2 parts coarse quartzitic gravel 2-3 mm in diameter and 1 part fern fibre or peat moss.

To pot up: Make sure the ingredients are thoroughly mixed. Place the plant in the centre of the pot and fill-in with mixture around the plant, making sure the mixture covers the root collar of the plant. Do not push or squeeze the mixture around the plant. Once fully potted up, simply knock the base of the pot twice firmly on a hard surface to stabilize the plant. The most appropriate watering method for this potting mixture is by overhead watering by hand or sprinkler.

This medium is suitable for areas where the climate is generally cooler and less evaporation takes place. The coarseness of the medium ensures excellent drainage, preventing the medium from becoming waterlogged.

A temperature-controlled, slow-release fertilizer such as Horticote 7:1:2 (27) can be added to the potting mixtures in both methods. In both methods, moss or algae may start growing on the surface of the medium after about 4 to 6 months, which often inhibits aeration and water penetration. This can be prevented by covering the surface area of the pot with quartz chips, coarse fern fibre or any other almost inert material that will block out sunlight onto the surface. In the case of the sphagnum moss medium, one often finds live sphagnum moss starting to grow. This is never a problem and is in fact a huge advantage. As the live moss is constantly taking up water, it evaporates, creating a cooling microclimate around the plant, which keeps the roots cool.

Re-potting

Re-potting ought to be done annually, preferably at the end of summer after the flowers have died and the new growing shoots have started to emerge. Should one miss this season, another appropriate time is at the end of winter or beginning of spring, as the plants start to show active growth.

Tip the pot over and gently shake the old mix loose. Remove dead roots and any other decaying matter such as the remainder of the previous year's tuberoid (the dead material is dark brown in colour and soft and structureless in texture). The plant is then immediately dropped into a basin containing a fungicide solution. Fungicides that are recommended are the wettable powder form of Dithane (active ingredient: mancozeb) or Kaptan (active ingredient: captab). The whole plant can be submersed in the solution. Rinse away all grit and other matter from the plant. This treatment has the advantage of preventing and treating fungal infections. Even allowing the plant to soak submersed for an hour will not damage it in any way. If the plant has produced any offsets which have matured, these can usually easily be separated during rinsing and potted-up separately. The cleaned plant can then be potted into a clean pot with new medium by holding the plant inside the new pot and placing the new growing mix between the roots. Fill the pot to the edge, making sure that the plant is potted up to the first rosette of leaves, and all white areas of the stem are covered.

Disas prefer to be slightly over-potted, and being gregarious by nature, enjoy being planted in a clump: three plants in a 6 cm diameter pot perform very well. As the plants are watered, the level of the potting mix will drop. Some granules of a

Above: *Disa tripetaloides* in habitat (see page 49)

Opposite: Newly re-potted *Disa* plant with the pot filled to the edge with growing medium slow-release fertilizer such as Osmocote or Horticote can then be pushed into the grit (maximum of 10 granules per pot). In the case of the granular medium, drench the plant well after re-potting so that water drains freely from the pot.

Watering

Whenever possible, watering should take place early in the morning when the water is at its coldest. Watering needs will vary depending on the season. During summer, water every second day (in some areas it will even be necessary to water every day). In winter, water once a week or even only every ten days during the colder months.

Although water quality is of utmost importance in all plants, disas in particular are not very tolerant of impure water. Even moderate salt levels are not suitable. Ordinary tap-water that is not too highly chlorinated can be used, although rain-water is preferred. To be

on the safe side, tap-water should be allowed to stand for 2 to 3 hours in order for the chlorine to evaporate. The pH value of the water should range between 5 and 6.8. Regulating the watering times is just as crucial. Although one likes to think that one is simulating the disa's natural environment as much as possible, growing a plant in a pot affects the microclimate around the roots dramatically, and although the drainage may be excellent, it is still an enclosed environment. The general rule of thumb when growing disas artificially is to increase watering during the warmer spring and summer seasons, and decrease watering during the colder autumn and winter seasons. Many a disa has succumbed to overly wet conditions during the colder seasons. There are three watering methods that have proved successful for growing disas: flooding, circulation and overhead watering.

Opposite: Overhead watering using a hand-held hose is best done early in the morning. Here John Cupido waters the Kirstenbosch Nursery *Disa* collection

Below: A water pH reading taken in a Table Mountain stream, indicating a highly acid value of 3.55

Flooding method
(also known as 'ebb and flow')

Add water to a deep tray in which the potted plants stand so that it reaches to approximately 2 cm below the rim of the pot. The water is then allowed to drain after a suitable time period, (from 1 to 4 hours). The drained water can then be used elsewhere. In summer the plants can be watered every day, while during the colder autumn and winter seasons, once per week or once every 10 days is usually more than enough. It is wiser to keep the plants on the dry side during the colder seasons.

The advantages of this method are:

1. As the water rises in the tray, only the roots are watered and the leaves of the plants are kept dry. The possibility of burn marks on the leaves and crown rot is thereby diminished.

2. The water supply is constantly fresh and drains away so there is no build-up of organic salts in the pot, which can occur if the water is re-circulated.

3. There is no algal growth in the water surrounding the plants.

Circulation method

In some instances, growers re-circulate the drained water, thereby ensuring an uninterrupted water supply. The water is kept at a depth of a third to half the height of the pot. Storage tanks, a reliable circulation pump and necessary plumbing are required. Care must be taken to ensure that the water temperature is kept at 15 °C or less, and the pH below 6.5. The water must be well aerated, with minimal salt build-up in the water from nutrients leaching out of the pots. The potting medium must allow for adequate upward capillary action of the water to ensure water supply to the upper levels of the pot.

Overhead watering method

This method uses a sprinkler system, a hand-held hose or a watering can to water the plants (see page 40). Its disadvantage is that all parts of the plant are watered, so if it is done at the wrong time of day (i.e. not early in the morning) it can lead to burn marks appearing on the leaves and rot developing in the axils of the leaves. The main advantage of this method is that the potting medium is constantly leached of excess salts. Care must also be taken to ensure that the potting medium retains sufficient moisture.

A combination of watering methods can also be utilized, for example, in summer the flooding method will keep water off the leaves, and reduce the chance of leaf burn and crown rot. In autumn, winter and spring, the overhead watering method is best. Watering once or, if necessary, twice per week will ensure adequate water supply for the plant, and leach out any salt build-up from the medium.

Opposite: Although most evergreen disas occur in nutrient poor soils (such as *Disa uniflora*), under cultivation they benefit greatly from feeding applied at the appropriate time (see page 43)

Feeding

In setting up a feeding programme for evergreen disas, it is important to bear in mind the growth cycle of the plants. Early spring marks the beginning of the growing season. As temperatures and light intensity increase, the rate of photosynthesis increases, thereby increasing vegetative growth. Thus the greater the light intensity, the greater the amount of nutrients the plant is able to utilize. Additional feeding is often desirable at this point. The plant should be mature at this stage and have a robust, rosette-like growth. The flowering spike is initiated and shortly afterwards a small new tuberoid will appear. Once the original plant has flowered, the flower spike turns yellow and withers. Any excess nutrients are re-absorbed into the newly developing tuberoid, which should by this time be of substantial size. From this new tuberoid a new shoot will appear and mature into a new plant and flower spike for the next season.

Two important issues should be remembered: feeding should be stopped once the flower has wilted, and re-potting should only occur after the flower spike is completely dead and gives way when gently tugged, and the new shoot has grown to approximately 3 cm in height.

Although disas can utilize varying strengths of fertilizer, it is always a good idea to start the feeding programme at half strength. Excessive feeding is never wise and can cause the plant to fail to produce a new tuberoid; with no sustenance, no new plant or flower will develop for the new season. Excessive amounts of nitrogen also cause soft, lax growth, and flower spikes will then need additional support in the form of stakes. The general tendency is to steer away from organic fertilizers partly because their nutrient concentrations aren't always exact, but more importantly because they can provide a source of pathogens in the form of bacteria and fungi. However good results have been achieved with

Percy Sargeant

certain organic fertilizers and these should not be considered as unfit.

Although not a fertilizer, Kelpak stimulates growth and root development. It is made from kelp and contains natural auxins and cytokinins, two of the most important growth hormones which stimulate shoot and root growth. It can be applied at the recommended dosage by watering can or mixed nozzle. A good standby fertilizer to use is a temperature controlled, slow release fertilizer. Several brands are available, amongst which is Horticote 7:1:2 (27), type 360. Type 360 refers to the period over which 80% of the nutrients are released, i.e. 360 days. Applying this type of fertilizer during the re-potting session will ensure a constant nutrient supply throughout the year.

Other fertilizers that can be used include the inorganic ones, Chemicult hydroponic fertilizer, Multifeed 10 and Phostrogen, and the organic Seagro and Nitrosol. Whatever fertilizer is used, bear in mind that at all times the leaching of excess nutrients must be possible. It is good practice to water with nutrient-free water several times between feeding sessions. This applies to flooding, circulation and overhead watering methods. An application of trace elements (Trelmix) once a month can be applied as foliar feed during the cooler times of the day, throughout the year.

Species to cultivate

The following evergreen *Disa* species are the most commonly grown and have also been found suitable for hybridizing.

Disa uniflora, pride of Table Mountain, red disa

The most popular and best known of all the disas, with the largest flower of all the southern African orchids. Plants can vary in growth habit from short-stemmed, robust individuals to those with flowering stems of up to 1 m long. Its distribution extends from the Cederberg Mountains in the west, to Betty's Bay in the south, to Riviersonderend in the east. Different colour forms occur in different mountain ranges: a carmine colour occurs on Table Mountain, a clear orange predominates in the southern regions and a deep orange-red in the Cederberg Mountains. Eastwards we find a clear pink *D. uniflora*. A yellow form, devoid of any red or pink, occurs sometimes in nature as a genetic mutation (see the yellow *Disa uniflora* story on page 48). A completely white or albino form was found on Table Mountain many years ago, although its exact location is unknown. *Disa uniflora* normally has a red or deep orange flower and is pollinated by the mountain pride butterfly, *Aeropetes tulbaghia*, that feeds on nectar contained in the spur (see photos on page 80). This butterfly is only attracted to colours in the red section of the spectrum, including the bright red flowers of the succulent *Crassula coccinea* and the red forms of the Guernsey lily, *Nerine sarniensis*. Thus a plant with a yellow flower (which is as rare an occurrence as an albino is in the animal kingdom) is ignored by the butterfly and its chances of being pollinated are virtually none. As a result, no seed can be produced and the plant's only other means of multiplication is by vegetative production of offshoots, a very slow procedure. Plants of the yellow form are often weaker than their more colourful counterparts, and easily perish when conditions are unfavourable. This is why the yellow *Disa uniflora* is so rare.

Right: The pink form of *Disa uniflora* from the Hex River Mountains

Below: The red form of *Disa uniflora* from Table Mountain

Dave van der Merwe

Wouter van Warmelo

ATLANTIC OCEAN

Cederberg

Hex River

Hex River

Franschhoek

Table Mountain

Table Mountain

Betty's Bay

Olifants

Cederbe

Oliants R.

Piketberg

Porterville

Berg

Paardeberg

Tulba

Malmesbury

CAPE TOWN

Table
Mountain

Paarl
Mountain

Paa

Kle

Wemmershoek

Stellenbosch

Franschhoek

Cape
Flats

False Bay

Hottentots Holland
Mountains

Palmiet

Cape of Good Hope

Cape
Hangklip

Betty's Bay

S a n d v e l d

Disa uniflora geographical forms
(see page 44)

GREAT KAROO

...eld Mountains
...Bokkeveld

Matjiesfontein ○

○ Laingsburg

Witteberg

Klein Swartberg

Seven
Weeks
Poort

G

...ex River Mountains

Anysberg

Ladismith

L i t t l e   K a r o

Rabiesberg

Koo Mountains

Worcester

Montagu

L a n g e b e r g

A

Robertson

Breede

Gouritz

...lliersdorp

Jonaskop ○

McGregor ○

Riviersonderend Mountains

Tradouw
Pass

Garcia's Pass

○ Riversdale

Swellendam

○ Albertinia

Greyton ○

Riviersonderend

Caledon

Potberg

○ Still Bay

...River Mount...

Cape Infanta

Bredasdorp ○

Elim ○

Cape Agulhas

INDIAN OCEAN

Map reproduced by kind permission of Struik Publishers
from *South Africa's Proteaceae* by Marie Vogts (1989)

The yellow *Disa uniflora* story

Extracts from articles by Hildegard Crous that appeared in the *South African Orchid Journal* (1997) and in *Veld & Flora* (1998):

"It is in the hot summer months, December through to the end of February, that one might come across an unusually vibrantly coloured flower growing in the coolness of a stream or splash area of a waterfall. *Disa uniflora* or pride of Table Mountain as it is also known, is a terrestrial orchid that occurs on Table Mountain and other Western Cape mountain ranges. The flowers, which are the largest of the approximately 93 *Disa* species that occur in the Western Cape, have various colour forms that range from a deep carmine red on the western ranges, orange in the southern ranges and pink in the more easterly ranges. However, on rare occasions a genetic variation leads to a pure yellow form making its appearance. This yellow form is devoid of any pink or red pigment throughout the entire flower. It is this unique feature that makes this rare occurrence so sought-after by breeders and collectors. This has also contributed to the disappearance of known yellow-flowering plants from the wild. The normally poor vigour of the yellow form can cause great difficulty in cultivating the plant under nursery conditions. Plants are more susceptible to fungal infections and show poor growth. As the yellow form is usually a singular occurrence, self-pollination is the only means of producing seed. Self-pollination however generally results in much less variation and less vigour, the latter not being desired. Cross-pollinating with another *D. uniflora* that shows less red pigment results in the seedlings flowering in various shades of pink. None of the seedlings are devoid of the red carotenoid pigment. The only solution to maintain a source of pure yellow genetic material is to cross pollinate with another yellow form of *D. uniflora*."

Disa tripetaloides

The most widespread species of all the evergreen disas, this robust, prolific flowerer is probably the easiest evergreen species to cultivate. It is amongst the easiest species to hybridize in the genus and is prominent in many breeding lines. Interestingly, although it is so widespread, it does not occur on the Cape Peninsula mountain chain. *D. tripetaloides* marks the beginning of the main flowering season, with flowers that appear in November. It is restricted to Table Mountain Sandstone habitats and two growth forms occur - the winter rainfall form that occurs from the Western Cape to the southern part of the Eastern Cape and flowers from November to January, and the summer rainfall form that occurs in the former Transkei of the Eastern Cape and southern part of KwaZulu-Natal, flowering from June to September. The summer rainfall form does not experience winters as cold as those in the Western Cape and starts flowering five months earlier.

When used in breeding programmes, it affects any resultant progeny by inducing an earlier flowering season. Plant height reaches up to 600 mm and flower colour and shape varies remarkably, from small speckled flowers to large, open, mostly white flowers. A rare pink form is seen now and then, and a very rare yellow form occurs in the Transkei. The number of flowers per stem varies up to twenty, and flowers can be well spaced along the stem or clustered closely together.

Below: *Disa tripetaloides* is probably the easiest evergreen *Disa* to cultivate

Percy Sargeant

Disa aurata

This little-known, yellow-flowering disa occurs in a very limited area of the Langeberg range near Swellendam in the southern Cape where it flowers from December to January. It was initially known as *D. tripetaloides* var. *aurata*, but was later upgraded to subspecies level as subsp. *aurata*, and was finally recognized as a species of its own on account of its longer lateral sepals, shorter spurs and bright yellow flowers. It never produces much seed in habitat as the plants are often under water due to excessive precipitation from the 'tablecloth effect' (see page 19), and many of the flowers are simply washed off or decay. However, like *D. cardinalis* and *D. tripetaloides*, the plant reproduces vigorously by the production of a multitude of stolons, which in turn send up shoots, and large mats of these disas can often be found in habitat. Plants reach up to 600 mm high but it has not been much used in hybridizing, probably due to its limited availability.

Disa cardinalis

This highly ornamental species is the toughest of all the evergreen disas and has a restricted distribution along stream banks of mountain slopes in the Riversdale district, flowering from October to December. It grows at lower altitudes than most of the other evergreen disas, is mostly found on north-facing slopes exposed to extreme temperatures and is hardly affected by thrips attacks during the hot, dry summers experienced in the winter rainfall parts of South Africa. Flower shapes vary greatly within the species, from those with well-rounded sepals, to the better-known forms with long drooping sepals. Sepal colour also varies in shades of brilliant red to crystalline orange to a pale orange-yellow. This clump-forming disa grows up to 600 mm high and reproduces vigorously by means of stolons. It has many attributes that make it desirable for breeding programmes, yet curiously this has not yet occurred to a large degree. The well-known hybrid *Disa* Kirstenbosch Pride is a primary hybrid of *D. uniflora* and *D. cardinalis*, selections of which include the deep pink 'Clare' and the dark red 'Meg's Delight'.

Left: *Disa aurata*

Opposite: *Disa cardinalis*

Percy Sargeant

Disa caulescens

A petite, white-flowered species with dark maroon barring on the median sepal, *D. caulescens* occurs almost exclusively in the western half of the Western Cape, in semi-shade along stream banks. It grows up to 400 mm high and occurs on Table Mountain Sandstone, forming dense mats as a result of the vigorous production of stolons. It flowers from November to January and is rare in cultivation.

Above: *Disa caulescens*

Opposite: The deciduous *Disa racemosa* (see page 72)

Disa Veitchii was the earliest artificial *Disa* hybrid, a cross between *D. racemosa* and *D. tripetaloides*, raised in Britain in 1891

Artificial hybrids

A hybrid is the resultant progeny of two species that have been cross-pollinated either naturally or artificially. Hybrids usually share properties from both parents and often have more vigour. The earliest man-made *Disa* hybrid was *Disa* Veitchii, a cross between *D. racemosa* (pod parent) and *D. uniflora* (pollen parent), and dates back as early as 1891.

A selection of some of the best-known artificial *Disa* hybrids

Hybrid Name	Parentage	Original Source
Disa Veitchii	= *D. racemosa* x *D. uniflora*	1891 - Veitch
Disa Kewensis	= *D. uniflora* x *D. tripetaloides*	1893 - Royal Botanic Gardens, Kew
Disa Premier	= *D. tripetaloides* x *D.* Veitchii	1893 - Royal Botanic Gardens, Kew
Disa Langleyensis	= *D. racemosa* x *D. tripetaloides*	1894 - RB Gardens, Kew & Veitch
Disa Diores	= *D. uniflora* x *D.* Veitchii	1898 - Veitch
Disa Watsonii	= *D.* Kewensis x *D. uniflora*	1900 - Royal Botanic Gardens, Kew
Disa Luna	= *D. racemosa* x *D.* Veitchii	1902 - Veitch
Disa Elwesii	= *D.* Kewensis x *D.* Veitchii	1903 - Elwes
Disa Blackii	= *D.* Luna x *D. uniflora*	1915 - Flory & Black
Disa Italia	= *D.* Blackii x *D. uniflora*	1918 - Flory & Black
Disa Julia A. Stuckey	= *D. uniflora* x *D.* Italia	1922 - Flory & Black
Disa Betty's Bay	= *D. uniflora* x *D.* Diores	1981 - van Niekerk
Disa Kirstenbosch Pride	= *D. uniflora* x *D. cardinalis*	1981 - Winter
Disa Helmut Meyer	= *D.* Kirstenbosch Pride x *D. uniflora*	1982 - Meyer
Disa Linda	= *D. caulescens* x *D. uniflora*	1982 - Meyer
Disa Kewbett	= *D.* Betty's Bay x *D.* Kewensis	1982 - van Niekerk
Disa Unilangley	= *D.* Langleyensis x *D. uniflora*	1983 - van Niekerk
Disa Kewpride	= *D.* Kewensis x *D.* Kirstenbosch Pride	1984 - Vogelpoel
Disa Unikewbett	= *D. uniflora* x *D.* Kewbett	1986 - Cywes
Disa Cardior	= *D. cardinalis* x *D.* Diores	1986 - Cywes

Opposite: The hybrid, *Disa* Helmut Meyer (*D.* Kirstenbosch Pride x *D. uniflora*) was raised by Mr Helmut Meyer in 1982 (see page 55)

Right, below left and right: Various forms of *Disa* Kewensis, a cross between *D. uniflora* and *D. tripetaloides*, originally raised at the Royal Botanic Gardens, Kew in 1893 (see page 55)

The hybrid, *Disa* Kirstenbosch Pride, a cross between *D. uniflora* and *D. cardinalis*, was raised by former Kirstenbosch Curator John Winter in 1981 (see pages 50 and 55)

Natural hybrids

The following natural *Disa* hybrids have been recorded:

Disa x *brendae*	=	*D. caulescens* x *D. uniflora*
Disa x *nuwebergensis*	=	*D. caulescens* x *D. tripetaloides*
Herscheliodisa x *vogelpoelii*	=	*D. graminifolia* x *D. ferruginea*

Top left: *Disa* x *brendae* (pink flowers, top left corner) and its parents *D. caulescens* (white) and *D. uniflora* (red)

Top right: Close-up of *Disa* x *brendae*

Above: *Disa* x *brendae* flowering in the Franschhoek Mountains

Deciduous disas

Winter-growing, summer-dormant disas

Many of these species require very specific conditions in order to grow. *Disa atricapilla, D. bivalvata, D. elegans, D. racemosa, D. tenuifolia* and *D. venosa* occur in acid soils in a variety of habitats including nutrient poor sandy loam, seasonal marshes on sandy flats, and seepage areas on mountain slopes. In summer these areas dry out, but the large amount of decomposed matter in the soil insulates the tuberoids from dehydration. These disas tend to flower most prolifically during the first summer after a fire. The following year may produce the odd flower, but generally, although the plant may produce leaves, no flowers are seen until after the next veld fire. Members of the genera *Bartholina, Holothrix* and *Pterygodium* are also found in this habitat.

Disa harveiana, D. longicornu, D. maculata, D. richardiana, D. rosea, D. virginalis and *D. vasselotti* prefer shallow fibrous pockets in vertical south-facing rock faces at high altitude. During winter a continuous supply of water dripping off the rock faces supplies the necessary moisture. During summer the 'tablecloth' cloud brought in by the south-easterly winds supply the minimum moisture needed to prevent desiccation of the tuberoids.

Disa barbata, D. cornuta, D. draconis D. ferruginea, D. graminifolia, D. hians, D. lugens, D. purpurascens and *D. spathulata* prefer drier conditions, some of which occur in areas with a high water table in winter. *D. ferruginea, D. graminifolia* and *D. purpurascens* occur on dry mountain slopes in acid sandstone, while *D. barbata, D. draconis* and *D. lugens* are found on acid, deep sandy flats. The rather variable and widespread *D. cornuta* grows in a variety of habitats including grassland mountain slopes in the summer rainfall areas and sandy flats in winter rainfall areas. *D. hians* is found in well drained, gravelly soils and the fairly widespread *D. spathulata* is found in both sandstone and shale soils.

Summer-growing, winter-dormant disas

Summer-growing, winter-dormant disas generally occur in heavier soils in dry or damp montane grassland, such as *D. crassicornis* and *D. woodii,* or in some instances in marshes or along streamsides, like *D. chrysostachya.*

Almost all the South African deciduous disas have long been regarded as highly fastidious subjects that can be grown quite easily for the first year or two following translocation from the wild, but inevitably succumb shortly thereafter. Maintaining these plants successfully under cultivation over an extended period depends on several factors. Firstly, their most important physiological characteristic is their symbiotic association with mycorrhizal fungi, in which certain fungi attach themselves to, and penetrate the roots of the orchid. Through this integrated unit of plant root and fungus, the plant, and in some cases also the fungus, benefits from the exchange of nutrients. Secondly, deciduous disas and other orchid genera such as *Bartholina* and *Holothrix* that have annually-replaced, single root-stem tuberoids as their rootstocks, have to re-establish the mycorrhizal symbiosis anew

Disa graminifolia (formerly *Herschelianthe graminifolia*) occurs on dry mountain slopes in acid sandstone (see page 68)

every growing season when new roots are formed, which provides at least a partial explanation for their more difficult cultivation compared with evergreen terrestrials.

Although both winter- and summer-growing disas are generally considered difficult to cultivate, the winter-growers are generally slightly easier to maintain over the longer term. For the purposes of the home gardener, increasing stocks of deciduous terrestrials by vegetative means is currently the only feasible method of propagation. However, truly successful cultivation of these orchids in the widest sense includes the ability to propagate them, in particular to propagate sexually and raise mature, flower-producing plants from seed.

Potting methods

A suitable aspect for the cultivation of most deciduous terrestrials is one receiving morning sun and afternoon shade, or bright light for as much of the day as possible. Due to the rather delicate nature of most species, the deciduous terrestrials are best grown in situations that are shielded from heavy rain and strong wind, such as on raised benches under protection of a fibreglass roof with open sides for good ventilation. It is of great importance to place containers in positions where they will not overheat on very hot days. In order to cultivate deciduous disas and other genera like *Bartholina* and *Holothrix* successfully from wild-collected plants, it is important to grow them in the soil in which they occur naturally, or at least to incorporate some soil from the natural habitat into the growing medium (see footnote below). [1]

1 It is illegal to collect South African orchids in the wild without a permit

It is of the utmost importance that the correct growing medium is used. The medium should not dry out too quickly, nor become compacted as it drains. In the bulb nursery at Kirstenbosch, a 2-3 cm layer of well decomposed acid compost (such as pine needle compost) is placed at the bottom of the container, and the rest of the container is then filled up with approximately three parts of coarse, washed river sand, and one part soil from the habitat or river sand mixed with some soil from the habitat. This method has been successfully used at Kirstenbosch to maintain winter-growing plants that are dependent on a symbiotic association with mycorrhizal fungi, such as *Disa spathulata* subsp. *spathulata* that has been maintained and vegetatively increased there for more than twenty years.

For summer-growing disas a slightly heavier medium is suggested, such as two parts coarse river sand or quartzitic grit, one part acid peat or fern fibre, and one part vermiculite, in addition to some soil from the habitat. A deep, 15 cm diam. terracotta pot has been found most suitable for delicate dwarf, winter-growing species like *Disa spathulata* and other dwarf terrestrials including *Bartholina burmanniana* and *Holothrix secunda*. For larger species such as the summer-growing *D. woodii* a plastic pot with a diameter of 20-25 cm can be used.

When planting, the tuberoids of deciduous disas must be handled with great care as they are rather soft and easily bruised, which often results in rotting. The tuberoids are planted so that the apex rests just below soil level. Tuberoids of winter-growing species are planted in autumn, while those of the summer-growers are planted in spring.

During the dormant period, tuberoids of both winter- and summer-growers should always be stored in dry soil and not be exposed to free air for any length of time as they soon desiccate and often succumb to attack by mealy bug, eventually perishing.

Watering and storage during dormancy

Most winter-growing disas produce leaves from late autumn to spring, and flower in late winter, spring and early summer. However certain winter-growers like *D. purpurascens* are proteranthous, producing their flowers in early summer, directly after the leaves have withered and died back completely. By contrast, other species like *D. ferruginea* are hysteranthous, producing their flowers in late summer or early autumn, just before new leaves appear. For all these species, regular watering should only start once the first leaf shoots appear towards the end of autumn.

The water used should preferably be slightly acidic or rain-water, or water that is free of dissolved salts as far as is possible. Tap-water should be allowed to stand for 2 to 3 hours to allow the chlorine to evaporate. Their growth pattern is very similar to that of the winter-growing bulbs of the Western Cape. During the active growing period a suggested watering procedure is a heavy drench applied once every 7 to 10 days, depending on weather conditions, allowing the medium to dry out almost completely before the next drench is applied. As soon as the foliage begins to wither as temperatures rise gradually from spring onwards, watering frequency should be decreased, allowing the growing medium to dry out completely during the summer dormant period, during which the containers should be placed in a cool, dry spot.

The summer-growing disas begin active growth in spring after the winter rest period. The growing medium should be kept on the dry side during winter and regular watering should only start once the new, leafy shoots appear in spring. As with the winter-growers, a heavy drench is suggested every 7 to 10 days during the spring and summer growing period, depending on weather conditions, gradually allowing the medium to dry out as the leaves wither and die back in late summer and early autumn.

Wouter van Warmelo

Disa ferruginea (see page 67)

Above: *Disa cornuta* requires an acid, well-drained sandy soil (see also back cover)

Opposite: *Disa bracteata* is easily cultivated and a recommended species for the beginner

Feeding

Fertilizers (applied at half the recommended rate) recommended for use with both winter- and summer-growers are Chemicult hydroponic fertilizer (inorganic) and Nitrosol (organic), and can be applied once per month during the active growing period. It is important not to fertilize more than once per month in order to ensure that any salt build-up is leached from the growing medium.

Winter-growing species to cultivate

Deciduous disas are seldom encountered in cultivation as they are much more difficult to maintain than the evergreen species, and are largely grown by specialist collectors. A number of species have however been successfully cultivated, although propagative material is unfortunately very difficult to obtain.

Disa bracteata
(formerly *Monadenia bracteata*)

Although not especially showy, this green, maroon and yellow-flowered plant is a good species for the beginner as it is a fast grower and easily cultivated, setting seed abundantly following self-pollination. It does not appear to be entirely dependent on an association with mycorrhizal fungi. A word of caution though, under ideal conditions it can become weedy, rapidly colonising adjacent containers and those some distance away in an orchid collection. It grows up to 300 mm high and occurs widely in the Western Cape and the southern parts of the Eastern Cape, flowering prolifically from spring to early summer.

Disa cornuta, golden orchid

This most attractive, robust species has purple and white flowers and is unusual in that it occurs in both winter and summer rainfall areas, its distribution extending from the west coast of the Western Cape to the Eastern Cape, KwaZulu-Natal, Mpumalanga and Lesotho, and further north into Zimbabwe. It grows up to 1 m high. In winter rainfall parts it grows on acid sandy flats and sandstone mountain slopes, while in the summer rainfall zone it is found in heavier soils in grassland. The winter rainfall form has been successfully cultivated for limited periods of time and is best grown in deep terracotta pots in an acid, well-drained sandy medium into which some soil from its habitat has been incorporated.

Cameron McMaster

65

Disa draconis, white disa (above)
A beautiful white-flowered species with purple markings that has a highly restricted distribution in the south-western coastal part of the Western Cape, occurring in deep acid sand. The plant grows up to 600 mm high and its leaves are proteranthous, the flowers appearing in late spring and early summer directly after the foliage has withered. It performs fairly well in cultivation in deep plastic pots in acid, very sandy soil. It is susceptible to over-watering, requiring the growing medium to be dried off almost completely before the next watering is applied, and a completely dry summer dormant period.

Disa ferruginea, red cluster disa (above)
This fiery-red or deep orange disa has been cultivated with some success but much has still to be learned regarding its exact requirements. It has an anomalous growth cycle in that if the tuberoid produces an inflorescence in the usual autumn-flowering period, then no leaves are produced in the ensuing winter-growing period. During the following autumn and winter, only leaves will be produced, and only in the following autumn will an inflorescence again be produced, under ideal conditions. The plant grows up to 500 mm high, is best grown in terracotta pots and requires high light intensity throughout the winter-growing period. It likes very acid, sandy soil, with the addition of some soil from its habitat incorporated into the growing medium.

Disa glandulosa

Successfully cultivated and flowered at the Royal Botanic Gardens, Kew, this disa is one of the miniatures. It is also self-pollinating, the pollen being deposited onto the stigma as the flower closes at the end of the flowering stage, shortly before wilting. It grows up to 200 mm high and has small translucent, pink flowers with a distinctive dark pink dot on each lateral sepal, close to the centre of the flower. Its foliage is quite distinctive as it is ever so slightly pubescent. It grows in mossy crevices on south-facing slopes and should be an ideal candidate for establishing in a 'planter tube' (see page 33). It has been hybridized successfully with *D. tripetaloides*, the pink translucent quality of the flowers being emphasized in the resultant flowers.

Disa graminifolia, blue disa, bloumoederkappie

A most attractive species that has violet-blue flowers with bright green petals in

Cameron McMaster

late summer and early autumn. It grows up to 1 m high and occurs amongst restios on coastal mountains from Cape Town to Port Elizabeth. It is an erratic flowerer: if an inflorescence is produced, then no leaves are produced in the ensuing winter-growing period. In the following late summer/early autumn seasons, an inflorescence is usually not produced, although leaves are. It is best grown in terracotta pots and is a difficult species to maintain in cultivation over an extended period, requiring some soil from its habitat incorporated into the growing medium.

Disa maculata

One of few blue-flowered disas, *D. maculata* grows in moss-filled rock pockets and ledges at relatively high altitude. The plants form a compact rosette, the leaves being quite short and somewhat succulent. The plants grow up to 300 mm high and the flower spikes are short and usually bear a single blue or mauve flower, often with a distinct green centre. The two lateral sepals are pointed at the tips and the median sepal has a short spur. Flowering takes place in October and November. For cultivation notes, see page 70.

Disa virginalis

This disa occupies the same type of habitat as *D. maculata*, and at first glance one could easily mistake it for a white form of *D. maculata*. It does, however differ quite significantly. The plants grow up to 300 mm high and the flowers are white, and display distinct pink-coloured lines or barring on the two lateral petals. The lateral petals are more pointed and concave in shape. For cultivation notes, see page 70.

Opposite: *Disa graminifolia* (see also pages 19 and 61)

Below: *Disa maculata* (blue) and *Disa virginalis* (white) (see also page iv)

Percy Sargeant

69

Percy Sargeant

Disa longicornu, drip disa (above)
Blue in colour, this disa is known for its very specific habitat preferences. It only grows on vertical south-facing rock ledges in rock fissures covered by sphagnum moss. The plant is constantly watered during the winter and spring months by a steady supply of run-off water dripping off the rock ledges. The plants grow up to 200 mm high and the tuberoids are often wedged deep into the rock fissures and are thereby protected from being dislodged should the mass of sphagnum moss drop to the ground. The plant differs from *D. maculata* in that it

has long, pendant leaves and the median sepal has a large spur with very distinct dark mauve venation.

Although *D. maculata*, *D. virginalis* and *D. longicornu* are not always found in exactly the same habitats, significant success has been achieved in cultivating them using the same method for all three. The plants are grown in a pocket of suitable medium, usually peat moss or sphagnum moss based, on a specially mounted, unglazed ceramic tile. The plants are planted in the moss and held in place by a strip of coarse shade cloth or mesh on the lower side of the tile.

On the top side of the tile a small water reservoir is made by positioning a strip of tile at an approximately 40° angle and secured in place by triangular sections of tile on each end. Leaving a narrow slit between the tile strip and the mounting tile ensures that when the 'reservoir' is filled with water, only a small amount of water will seep down the length of the tile and adequately drip/seep irrigate the plants at the lower end of the tile. The water flow can furthermore be regulated by lining the reservoir with a small piece of bidum cloth, thereby ensuring even slower seepage.

Another method, the 'planter tube', which has proven most successful, is perhaps easier to construct (see page 33). A length of bird netting, 1 m x 40 cm is sewn up down the long side and the bottom end. This 'tube' is then filled with sphagnum moss which has been soaked in water and had the excess water squeezed out. The moss should be firmly packed into this 'tube' with no empty spaces. Once filled, the excess netting at the open end is knotted. The tube can then be suspended from a strong bar in the greenhouse. The bottom end of the tube should rest gently on a tray or planter bench. Holes are cut into the bird netting at evenly spaced distances and cavities hollowed out sufficiently to accommodate the plants. Wrap each plant in a small amount of moss and wedge into the cavities. Winter-growing, summer dormant species should be planted in the top quarter of the planter tube. Evergreen disas can be planted lower down. In this way the winter-growing, summer dormant disas can be kept drier during the summer months.

Watering is done manually by placing a hose at the top of the tube and allowing the water to seep down through the 'tube' of sphagnum moss. Alternatively insert a dripper line into the top of the 'tube'. This is connected to an automatic watering system. A small regulator tap can be inserted in order to regulate the water flow.

In both these cultivation methods, the advantage is that over-watering is never a danger as the water is constantly draining away due to gravity, thereby simulating the drip effect. Salt build-up in the medium is never an issue and in the case of the 'planter tube', fertilizer (as for the evergreen disas, see page 43) can be applied during the growing season.

Disa lugens var. *lugens* (formerly *Herschelianthe lugens*), bloumoederkappie (see photo below)

An intriguing plant with a creamish-green, hooded median sepal, mauve lateral sepals, white petals and a prominent green, beard-like lip. Its distribution extends from Atlantis on the Cape West Coast to Grahamstown in the Eastern Cape. It grows up to 1 m high and occurs on deep, acid sandy flats, usually within the protection of thick clumps of the restio *Thamnochortus erectus*. It is becoming increasingly rare due to coastal housing developments. It has been successfully cultivated for a number of years in a deep container in soil from the habitat, within a clump of *Thamnochortus erectus*.

Cameron McMaster

Disa racemosa

This beautiful, yet seldom seen species has a fairly wide distribution in marshy areas of the Western Cape and southern parts of the Eastern Cape, flowering in November and December, following fires that occurred during the previous summer season. The plant grows up to 1 m high but it is a shy flowerer, except in the first season after a veld fire, during which it flowers prolifically, producing copious amounts of seed. However, hardly a single flower will appear again until such time that the next veld fire occurs. On investigation of the plants in habitat, large numbers of plants can be found, and it is most prolific, producing multiple stolons, each of which develops into a new plant. One of the earliest hybrids was *Disa* Veitchii, a cross between *D. racemosa* and *D. uniflora*, bred as early as 1891. When using *D. racemosa* in hybridizing programmes, it is wise to use it as the pollen parent and not the seed (pod) parent; if used as the seed parent, the reluctance to flower is sometimes carried over to the progeny.

Left: *Disa racemosa* (see also pages ii, 23, 53 and 55)

Opposite: *Disa sagittalis*

Disa sagittalis

An easily cultivated plant, this species is mostly found growing between sandstone boulders and in cracks on sandstone ledges, or occasionally along stream banks. It occurs in the southern parts of the Western Cape and the southern and north-eastern parts of the Eastern Cape, and the southernmost part of KwaZulu-Natal, growing in semi-shade in pockets of moss on stony well-drained soil. The plants grow up to 300 mm high and readily reproduce vegetatively by means of stolons, forming dense clumps under ideal conditions. Its white to pale mauve flowers are borne from spring to early summer and the plants undergo a distinct dormant period from midsummer to early winter, during which the growing medium should be kept completely dry. This species is best grown in terracotta pots.

Percy Sargeant

73

Disa spathulata (formerly *Herschelianthe spathulata*), oupa-met-sy-pyp (see below) This species grows up to 300 mm high and comprises two subspecies, the very variable subsp. *spathulata* that occurs in sandstone and shale soils in the western parts of the Western and Northern Cape, and subsp. *tripartita* that grows on shale in the extreme south-eastern part of the Western Cape. *D. spathulata* subsp. *spathulata* is a relatively long-lived and easy species in cultivation, multiplying vegetatively by the production of extra tuberoids under ideal conditions. It flowers in spring and early summer and is best grown in terracotta pots. It requires some soil from its natural habitat incorporated into the growing medium.

Disa tenuifolia
(below, formerly *Disa patens*)
(see also photo on page 101)
This charming little yellow disa is mostly
only seen after fire. It grows up to
300 mm high and its preferred habitat is
deep, loamy, acidic sand. It is often seen
near to *D. racemosa* and *D. bivalvata*
populations. Although not yet grown
successfully in nursery conditions, it has
been propagated and flowered *in vitro*.
Once the hardening-off process has
been mastered, this should prove to be a
delightful candidate for cultivation.

Disa venosa
(below, see also page 26)
This disa is not often seen and as it is so
similar to *D. racemosa*, the two are often
confused. It is restricted to the Western
Cape, occurring in marshy habitat on
Table Mountain Sandstone, flowering in
November at the same time of year as
D. racemosa. It grows up to 500 mm
high with pale to deep pink or rarely pure
white blooms, usually flowering following
late summer fires of the previous season.
D. venosa can be distinguished from
D. racemosa in the smaller stature of the
plants and flowers, and by the triangular
shape of the median sepal, in contrast to
the rounded, hood-like shape in the latter
species. It is rare in cultivation.

Percy Sargeant

Summer-growing species to cultivate

Propagative material is unfortunately very difficult to obtain, but the following species have been successfully grown:

Disa crassicornis

This beautiful, robust, sweet-scented species grows up to 1 m high and was first cultivated as early as 1879 at the Glasnevin Botanic Gardens in Scotland. It grows in grassland in a number of mountainous areas of the Eastern Cape, KwaZulu-Natal and Lesotho and is found from near sea level up to 2 700 m. It is unfortunately rather short-lived in cultivation and much has yet to be learnt regarding its exact requirements for long-term survival. It has unusual growth and flowering cycles; the tuberoid, which can last for longer than two years, produces a sterile shoot next to the fertile flower-producing shoot. The fertile shoot that appears in spring, grows throughout the summer but does not produce an inflorescence. It remains inactive during the cold winter months but begins growth again the following spring, and flowers in summer. The plant is best grown in deep terracotta pots in a well-drained, acid medium containing some soil from the habitat, and needs a dry winter rest.

Disa woodii

One of very few summer-growing disas that is easy to grow, this species has bright yellow flowers borne in a dense, sturdy upright spike from spring to early summer. Like its winter-growing counterpart *D. bracteata*, its flowers are self-pollinating and it has a tendency to become weedy in disturbed areas such as on road verges and embankments. It grows up to 700 mm high and requires sunny conditions when grown outdoors in temperate climates, or high light intensity for as much of the day as possible when grown under greenhouse conditions. It has a wide distribution extending from the summer rainfall parts of the Eastern Cape to KwaZulu-Natal, Swaziland, Mpumalanga, Gauteng and Limpopo, extending into Zimbabwe and Mozambique.

Cameron McMaster

Above: *Disa crassicornis*

Opposite: *Disa woodii* (see also pages 116–117)

Below: *Disa tripetaloides* is easily propagated by seed as well as by separation of tuberoid clumps (see page 49)

Opposite: *Disa uniflora*

Louis Vogelpoel

PROPAGATION

Vegetative propagation of evergreen, winter- and summer-growing disas (and other deciduous terrestrial orchids including *Bartholina* and *Holothrix*), whether by simple separation of rootstocks or by complicated *in vitro* techniques such as meristem culture, is of great benefit to orchid growers, but as far as species conservation is concerned it has limited value in preserving genetic variation. Truly successful propagation of orchid species can only be achieved if one can raise mature, flower-producing plants from seed. Most orchids are difficult to grow from seeds because they contain no food and have to form a symbiotic association with mycorrhizae in the soil to obtain nutrients. Fortunately the seeds of evergreen disas can be germinated relatively easily without using sterile techniques. However the seeds of deciduous species require complicated procedures, under sterile conditions (see page 90). The tiny dust-like seeds of deciduous disas from the winter and summer rainfall regions of South Africa are wind-dispersed, while the slightly larger seeds of the evergreen species *D. cardinalis*, *D. caulescens*, *D. tripetaloides* and *D. uniflora* are dispersed by running water (hydrochorous).

Pollination

In order to obtain seeds of *Disa* species under cultivation, it is necessary to learn how to pollinate the flowers by hand. Seeds can be obtained both by self-pollination (also known as inbreeding or 'selfing') and by cross-pollination. In self-pollination the stigma of a flower is pollinated with the pollinia from the same flower, while in cross-pollination the stigma is pollinated with pollinia from the flower from another plant. In large-flowered species like *D. uniflora* and its allies, selfing and cross-pollination procedures are quite easy and simply entail detaching the pollinia from the anther of the pollen (male) parent with the tip of a matchstick or toothpick and smearing them onto the stigmatic surface of the seed (female or pod) parent when the flowers are fully open (see figure on page 12).

Cross-pollination is the preferred method as it results in greater genetic variation in seedlings. Self-pollination is usually performed to bring out desirable recessive genes, but its non-beneficial effects usually include less vigorous offspring that are more prone to disease. If one has decided to perform cross-pollination, it is essential to emasculate the flower of the seed parent by removing the anther before hand-pollination is carried out, so that self-pollination of the flower cannot take place. If pollination has been successful, the flower rapidly fades within about 3 days and the capsule begins to develop. After 5 to 6 weeks, the oblong-shaped capsule ripens and begins to split lengthwise. Immediately as it begins to split, or ideally just before splitting commences, the capsule should be cut off just below its base and placed in a small open dish to allow it to dry out and release the mass of tiny, pale brown seeds.

Opposite above and below: *Disa uniflora* being pollinated by the mountain pride butterfly (*Aeropetes tulbaghia*).

This large, colourful butterfly is only attracted to colours in the red section of the spectrum; the pollen sticks to the legs of the butterfly and is carried from one flower to another, effecting pollination (see also page 44)

Evergreen disas

Seed propagation

Conventional sowing method

Evergreen disas produce relatively large seeds, which are designed to float on the water surface, enabling them to be dispersed downstream from where the parent plant is situated. Once trapped in a suitable site such as a pocket of sphagnum moss, the seeds will start to germinate after approximately 3 weeks. The young seedlings will remain secure throughout the wet winter season and grow during the following summer season. If all goes well, the seedlings will mature to flowering stage in their second or third year after germination.

Germinating evergreen *Disa* seeds at home is easily simulated by placing some sphagnum moss into a shallow seed tray and dispersing the seeds as evenly as possible onto the moss surface. It is advisable to first place the sphagnum moss in boiling water, then let it cool completely before sowing in order to prevent it from growing too vigorously and overwhelming the developing seedlings. Place the seed tray in a cool, protected area and ensure that the moss is kept moist at all times by watering with a fine rose. After germination has taken place after about 3 weeks, allow the

Above: Three-month-old seedlings of *Disa uniflora* raised by the conventional method

seedlings to reach about 1 cm in height before pricking them out and potting-up.

In vitro sowing method

This is by far the more complicated method and requires expensive equipment, however the rate of success achieved is most gratifying. After pollination it takes approximately 4 weeks for the seed pod (capsule) to mature. The seed is sown and grown in a tissue culture laboratory under aseptic, controlled conditions where the temperature, amount of light, heat and moisture are controlled. Three basic phases are followed to ready the plant for hardening-off and growing-on in the nursery to nurture the plant to maturity.

Phase 1: Seed sowing with the 'green pod method' is by far the most efficient method. The 'green pod' is picked shortly before the seed ripens, usually indicated by a yellowing of the seed pod (in *D. uniflora* this occurs 28 days after pollination). The pod is trimmed at both sides and any dead matter carefully scraped off. Dip the whole pod in 96% ethanol and then rinse in a 30% NaOCl (Sodium hypochlorite) solution to which a drop of detergent has been added. After rinsing for 10 minutes, place on a sterile piece of paper to dry in the Laminar Flow Bench which has been switched on and wiped down with a sterilant. Once the pod has dried after approximately 5 to 10 minutes, cut the pod open. Once cut open, tap the seed into a flask onto a pre-prepared agar medium in which *Disa* specific nutrients are suspended. The jar is then sealed and placed in a growth room where lighting is provided for 16 hours a day and temperature is kept constant at about 23–24 °C.

Above: Cutting open a green pod of *Disa uniflora*

Phase 2: Germination usually takes place after 3 to 4 weeks. The seed swells and develops into a protocorm, a round body from which the first root hairs and leaves appear. Once the seedlings have developed two leaves of approximately 3 mm long, the plantlets should be separated and transferred onto new medium, by which time most of the nutrients have been absorbed. It is important not to overcrowd the flasks as this will increase competition for the nutrients.

Phase 3: Once the plantlets have grown to about 4 cm in height, transfer them once more into a larger flask, placing only as many plants into the flask that will allow each plant adequate space to grow. This is the final phase, after allowing the plant to grow for another 2 to 3 months, roots should have developed and the plant increased dramatically in size. At a height of 5-10 cm the plantlets are ready to harden-off and grow-on in the nursery.

Opposite above: Tapping *Disa uniflora* seeds into a flask onto a pre-prepared agar medium containing *Disa* specific nutrients (see page 83)

Opposite below: Developing *Disa uniflora* protocorms (left flask), and young plantlets (right flask)

Above: Transferring *Disa uniflora* seedlings onto new medium

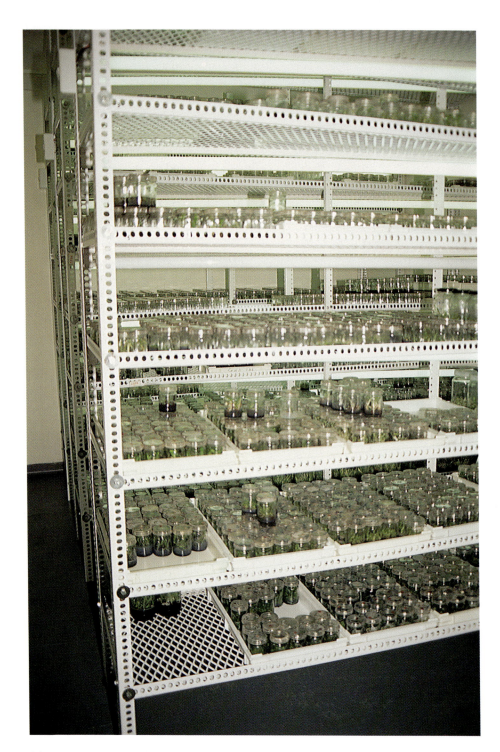

The whole process takes approximately 18 months, and first flowering usually takes place three years after germination.

Successful hardening-off

Unless one has a specifically designed greenhouse for hardening-off throughout the year, the best seasons to harden-off are in spring and early summer. Open the flask and add approximately 2-3 cm water [2]. Keep the open flask for two days in a cool, well ventilated area under a roof. Place a piece of garden fleece over the flask in order to reduce evaporation. If conditions are extremely hot, keep the flask indoors for one day and then outside for two days in a suitably protected area. Watch the level

of the water. If the rate of evaporation is excessive, top-up with water. After two days, remove the plants from the flask, rinse in a suitable fungicide such as Dithane M45, making sure that as much of the agar as possible is rinsed off the roots.

Pot-up in the sphagnum moss mixture recommended on page 34 in a shallow seedling tray or small pot. Shallow trays or pots are essential as the plants often do not have well developed roots at this stage, and it is critical that the roots receive adequate water. If conditions are extreme, cover the plants with garden fleece for another week. The upper half of some of the leaves might scorch and wilt, and eventually turn yellow, however there is no need for concern as the plant will in any event produce new leaves and discard the leaves that had formed in the flask. Once the roots are well developed the plant can be transferred to a deeper pot.

2   The amount of water is determined by the size of the plant: generally one should not add more than one third the height of the plant. If the plants are really large (more than 10 cm), water up to half the height of the plant would be better, as the greater leaf surface exposed will increase the amount of water evaporating from it.

Opposite: The sterile environment of a *Disa* tissue culture laboratory

Above: Newly separated *Disa* plants

In vitro, or tissue culture, may seem a very complicated and costly method of propagation. However, orchid seeds are very vulnerable to the elements as they have no endosperm protecting the embryo. This means that the embryo is simply covered by a thin outer layer, the seed coat. The seed thus needs very specific, unique conditions to germinate and grow successfully and safely into a plant. In vitro, which literally translates to 'in glass' is the method of cultivating plants inside a closed container where conditions are aseptic and free of pathogens. In this environment the temperature, moisture, lighting and nutrition is controlled to the maximum benefit of orchid seed germination and seedling growth. Once the seedlings have been nurtured to a manageable size, the plantlets can be more easily

adapted to nursery conditions, and made more resilient to external factors.

Propagating the yellow Disa uniflora

In 1991, a reasonably vigorous yellow-flowering form of Disa uniflora was collected by staff members of Kirstenbosch National Botanical Garden on the Table Mountain range. It was successfully cultivated and flowered in the Kirstenbosch nursery, and pollen was harvested and stored. Fortunately, a year later another yellow-flowered D. uniflora was found in the Hex River mountain ranges and brought into cultivation (see pages 46–47). When both forms flowered in February 1993, the flowers were cross-pollinated by hand. The resultant seeds were sown and nurtured in vitro until the seedlings were strong

enough to grow under normal nursery conditions. In this instance, as often happens when hybridizing, the seedlings were very much more vigorous than the parent plants. In January 1997 the first generation intraspecific hybrid seedlings flowered. All the plants had retained the pure yellow colour with no trace of the red anthocyanin pigment. From these forty-odd plants, selections to improve form and colour were made. One clone, 'Kirstenbosch Supreme', was awarded an Award of Merit by the Disa Orchid Society of South Africa (AM/DOSA). Whereas both parent plants had somewhat reflexed and unevenly shaped flowers, the resultant progeny had a more uniform shaped flower, more even colouring and well shaped sepals. Unfortunately the improved vigour is not discernible in photographs. This first generation intraspecific hybrid will in turn be hand-pollinated and the resultant seedlings made available, as this colour form is very much sought after.

Vegetative propagation

Vegetative propagation of evergreen disas is best carried out in late summer and early autumn as the new leaf shoots start to develop. These new shoots appear close to the mature plants and develop at the tips of stolons produced by the mature plant. If the new plants are healthy and well established, they can be divided by immersing the whole pot into water in order to loosen the potting medium from the roots. The dying mature plant can be discarded and the individual new plants should then be carefully separated and each potted-up into new potting medium. Depending on size, one to several new plants can be placed into a pot, in which they can establish themselves until flowering size has been reached. Extreme care must be exercised when handling the roots and tuberoids of the plants as they are brittle and easily damaged.

Opposite: *Disa uniflora* 'Kirstenbosch Supreme', a selection from a cross between yellow forms from Table Mountain and the Hex River Mountains

Left: *Disa uniflora* plant with stolon and developing new plant at its tip

Deciduous disas

Seed propagation

The flowers of deciduous disas are pollinated in the same way as those of the evergreen species but as the flowers of most of the deciduous species are much smaller, hand-pollination is more difficult, and the seed pods and seeds are much smaller.

Deciduous winter- and summer-growing disas cannot be raised successfully from seeds sown under non-sterile conditions (as one would normally sow a tray of *D. uniflora* seeds, for example) because the seedlings are dependent for survival on a symbiotic association with a specific fungus. In addition, further requirements for germination of viable seeds, such as sufficient moisture, optimum temperature and light levels, also have to be present. The fungus penetrates the roots of the orchid seedling and through the exchange of nutrients, nutritional benefit is obtained by the seedling, and in some instances, by the fungus as well. It is especially during seed germination and subsequent growth of the seedling that the orchid's dependence on its associated fungus is greatest. Growing and propagating these species from seed requires knowledge of the mycorrhizal associations which these orchids form, and is dependent on laboratory research at scientific institutions.

Generally, two methods are used for germinating seeds of deciduous disas, the symbiotic and asymbiotic methods, both of which are complicated procedures and take place *in vitro,* under sterile conditions. Although both methods are essentially beyond the means of the amateur enthusiast, they are briefly summarised below. Detailed information on these methods is available in the Orchidaceae chapter in the publication *Bulbous plants of southern Africa* (Du Plessis and Duncan 1989).

Symbiotic method

In a nutshell, in the symbiotic method the orchid seed is germinated in the presence of an appropriate mycorrhizal fungus, in a medium from which the fungus can draw nutrients that are transformed and passed on to the orchid. The first step in the symbiotic method is to isolate the mycorrhizal fungus from the roots of the orchid and store it in a culture medium. The orchid seeds are then sown on a suitable medium, which is inoculated with the appropriate fungus. Once the seedlings have grown to a suitable size, they are lifted together with their mycorrhizal fungi, and potted-up in a soil-based medium.

Asymbiotic method

In the asymbiotic method, the seeds are sown in a medium in which the essential nutrients are supplied in a formulation that the orchid seed can utilise directly. Unfortunately, obtaining nutrient formulations suitable for deciduous terrestrials is problematic as they usually fail to germinate in well-known formulations like Knudson C (that have been successfully used in germinating seeds of epiphytic orchids) due to the essential minerals being supplied in the form of inorganic salts. However, several South African terrestrial species have been successfully germinated using formulations in which the nutrients are supplied in organic form. A further difficulty experienced with the asymbiotic method is the process of transplanting seedlings from the sterile nutrient medium to a soil-based medium, as they are effectively cut off from their nutrient supply once planted into the soil-based medium. This process should ideally be achieved before the young plants enter dormancy because if they enter dormancy in the nutrient medium, the young tuberoids usually fail to sprout in the following growing season. In order to establish the mutualistic association between the young seedling and the appropriate mycorrhizal fungus once the seedlings have been planted, the soil-based medium should contain some soil taken from the habitat of the plant, that hopefully contains the appropriate fungus.

The following yeast extract formula has been successfully used in germinating seeds of deciduous disas and other terrestrials. In this formula, the yeast extract is the major source of nutrient.

Yeast extract formula

2 g yeast extract
6 g agar-agar
10 g sugar
4 g peptone
1000 ml distilled water

The medium is made up by heating the distilled water and dissolving the solids in it. The mixture is then poured to a depth of 1 to 2 centimetres into the containers in which the seed is to be sown, such as Erlenmeyer flasks or petri dishes. These containers and their contents are then sterilised in a pressure cooker or autoclave for 20 minutes at 103 kPa, after which they are left to cool overnight. The seed pod is then sterilised (see phase 1 on page 83) and the seeds are tapped onto the surface of the germinating

medium. The flasks are then sealed and placed in a growth room.

The seeds of the deciduous winter growers remain dormant in summer and will only germinate during autumn after temperatures have dropped. When the seedlings are ready to be transplanted into a soil-based medium, the nutrient medium must be cleaned off the seedlings as far as possible to prevent the development of parasitic fungi and bacteria. Throughout all these above processes, a sterile environment must be maintained at all times.

Percy Sargeant

Above: *Disa harveiana* subsp. *harveiana*

Vegetative propagation

For the home gardener or keen enthusiast, vegetative propagation is at present the only feasible means of increasing stocks of deciduous disas. Tuberoid clumps of deciduous disas and other winter-growing terrestrials like *Bartholina burmanniana*, *Holothrix secunda*, *Pterygodium alatum* and *Satyrium carneum* are best separated in late summer (mid to late March in the southern hemisphere) just before active vegetative growth begins. For deciduous summer-growing species such as *Disa crassicornis* and *D. woodii*, the clumps should be separated as temperatures rise in early spring (early August in the southern hemisphere). The rate of additional tuberoid production varies from one species to another, and additional tuberoids are not necessarily produced by each mature plant every year. Reaching the stage where

vegetative propagation can take place is a lengthy process, and separation of tuberoids should only take place once the daughter plants have become fully established. For this purpose it is suggested that tuberoid clumps be separated approximately every 3 to 5 years, depending on the species.

The winter-growing *Disa spathulata* subsp. *spathulata* has been successfully vegetatively propagated in the Kirstenbosch bulb nursery for more than 20 years; a single plant collected in the wild in 1985 has, through simple separation of tuberoid clumps, been increased to more than 20 plants by the year 2006. The rootstocks of all deciduous disas must be handled with utmost care as they are easily bruised against hard surfaces, the wounds providing an easy entry point for fungal infection. Separated rootstocks should be re-planted immediately, as the delicate

tuberoids are susceptible to desiccation when stored outside of soil media for any length of time. When re-planting tuberoid clumps of *Disa, Bartholina, Holothrix, Pterygodium* and most other deciduous terrestrials it is essential that some soil from the pot in which they have been growing is incorporated into the soil medium of any additional pots into which

daughter plants are planted, in order to ensure that the mycorrhizal association is maintained.

Tuberoid clumps of deciduous, winter-growing disas like *Disa spathulata* (opposite) and *Bartholina burmanniana* (below) are best separated in late summer

The Disa barbata Project

An article by Benny Bytebier that appeared in Plant Talk *journal in 2004.*

Disa barbata is undoubtedly one of South Africa's most threatened orchids. This pretty, white-flowered plant with a frilly lip is known in Afrikaans as 'ou man-met-sy-baard', which translates to 'old man with his beard'. It can only be spotted in the field during a two week period around the end of October when it is in flower, since vegetatively it resembles and blends in with the Cape reeds (Restionaceae) among which it grows. Herbarium specimens dating back to the late 19th century show that this orchid was once found in the sandy flats around Cape Town in areas like Kenilworth, Rondebosch and Pinelands. Unfortunately all of these are now suburbs in the concrete jungle of a sprawling metropolis, and the last sighting in Cape Town dates back to 1950.

Only one known population of this orchid remains. This is close to Malmesbury, about 50 km north of Cape Town. Here it is protected in a nature reserve together with several other species endemic to the coastal, sandveld fynbos. Although the area is heavily infested with the Australian Port Jackson willow (*Acacia saligna*), over the last couple of years the reserve itself has been cleared of this invasive tree species under the Working for Water Programme, which aims to conserve the water resources of South Africa by removing alien vegetation.

In 2001, while I was collecting fresh material for a phylogenetic analysis of the whole genus, I detected only eight plants after a day-long search in the reserve. I reported this alarming news to the management authority and together we drew up a rescue plan, which received financial support from the Endangered Species Fund of the Chicago Zoological Society and the Chicago Board of Trade.

Our approach to saving this species is three-pronged and involves three partners: the Western Cape Nature Conservation Board, which is the management authority for the reserve, the University of Stellenbosch and the Cape Institute of Micropropagation, a commercial tissue culture laboratory. First and foremost, we are hoping to boost the current population size by propagating plants through *in vitro* seed germination. Secondly, we are surveying the reserve as well as the surrounding area to see if there are any more populations. Thirdly, we are investigating the ecology and reproductive biology of this species. Orchid seeds are dispersed by wind and are extremely small. They require an association with mycorrhizal fungi for germination, and therefore the recruitment of new plants in nature is generally low. Given the right conditions orchid seeds can be germinated *in vitro* highly efficiently. With this in mind, we engaged a tissue culture company with experience in orchid tissue culture and a heart for conservation to produce new plantlets.

Although there are many technical hurdles to overcome, like seed dormancy, we have so far managed to germinate seeds and raise them to the protocorm stage. Once the plantlets have grown sufficiently they will be weaned and re-introduced into the reserve to increase the population and thus prevent the species from a possible random extinction event.

The genus *Disa* has horticultural appeal. Our plan is therefore to sell some of the artificially propagated plants to interested hobbyists. This will help prevent possible poaching of specimens from the reserve and will recover some of the costs we incurred in optimising the tissue culture techniques for this species.

Also, field surveys over two concurrent years have given us a better indication of the true size of the population. Each flowering plant was marked and its co-ordinates recorded using a GPS device. Within the reserve we can now report that there are at least 65 individuals. We also discovered a second sub-population just outside of the reserve, and this consists of at least 50 plants. While not high in terms of numbers, it is better than was previously thought. Our field observations also show that although the plants are perennial, they do not flower every year. Out of 54 plants observed in flower in the reserve in 2002, only four were seen to flower in 2003. At the same time 12 'new' plants appeared in 2003. Unfortunately the pollination success is generally low. Only 15% of the flowers observed in 2002 reached the stage of a mature seed pod. All other flowers were either not pollinated or were eaten, possibly by herbivores. The long-term survival of this species is of course critically dependent on the continued presence and conservation of

its pollinator, which we identified during this field season as *Xylocopa rufitarsis*, a species of carpenter bee. These bees depend on dead wood for nesting sites and this means that fire, an integral part of the fynbos ecosystem, will have to be managed carefully in the future.

With 9 000 species packed into 90 000 km^2 and a 69% endemism rate at the species level, the Cape Floristic Region harbours many endangered plants. *Disa barbata* is just one of them, but like any other, it is worth fighting for.

Benny Bytebier is an East African botanist who is currently at the University of Stellenbosch, South Africa.

William Liltved

Below: A caterpillar feeding on the leaves of *Disa uniflora*

Opposite: *Disa cardinalis*

PESTS AND DISEASES

Pests

Aphids or greenfly as they are also known are small, soft sucking insects that may be black or green. They infest developing *Disa* flower buds, especially during hot dry weather, resulting in deformed flowers. They also occur on the undersides of leaves, and the honeydew they secrete attracts ants. Aphids can be controlled with tetramethrin (e.g. the aerosol Garden Gun) for small plant collections. Hold the aerosol can at least 30 cm from the plant when spraying, as the intense coldness and the carrier material can cause leaf burn. For large plant collections, a wettable powder insecticide such as pirimor (e.g. Pirimicarb) works best.

Caterpillars of various kinds including boll-worm are partial to the leaves of evergreen *Disa* species and particularly their flower buds. Ideally the solution would be to use a systemic pesticide, however these are known to affect disas adversely. During early summer when the flower buds are developing, a regular spray with a contact pesticide such as carbaryl (e.g. Karbaspray) should control the problem. The adage 'prevention is better than cure' is very apt when growing disas. A constant watch should

best to control the small yellow larvae by spraying preventatively with fenthion (e.g. Lebaycid) at monthly intervals, from early to late summer.

Red spider mites only become a problem on evergreen *Disa* plants if the environment is hot and dry, and are an indication that the growing environment is unsuitable. Provide a cool, moister environment and this pest will stay away. These tiny red arachnids occur in large numbers on the undersides of leaves and on flowers, resulting in a silvery web-like covering. They cause desiccation of the tissues and in severe cases, death of the affected parts. They can be controlled with chlorpyrifos (e.g. Chlorpirifos) or mecaptothion (e.g. Malasol). Mineral oils like Oleum should not be used on orchids with soft, fleshy leaves like *Disa, Holothrix* and *Satyrium.*

be kept on the plants and action taken when the first affected plant is noticed.

Gall midge fly larvae are prevalent in summer and cause damage to developing *Disa* flower buds and flower stems, and also result in secondary bacterial infection and collapse of affected tissue. In susceptible areas it is

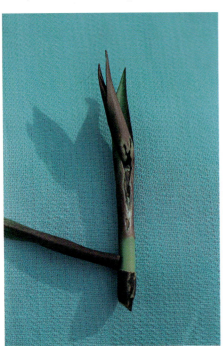

Snails and slugs are very partial to young *Disa* shoots, and can be controlled by scattering iron phosphate pellets (e.g. Biogrow Ferramol) around the outside of the pots and under raised benches. This is a natural snail bait and is safe around children and pets.

Thrips are tiny, elongated, brownish-black insects that occur in large numbers, rasping the surfaces of flower buds and the undersides of *Disa* leaves. They are identified by the silvery streaks that result in deformed flowers and leaves. They are prevalent in hot dry weather and are an indication that the growing environment is unsuitable for evergreen disas. By providing a cool, moister environment for evergreen species, thrips will stay away. In highly susceptible areas they should be controlled preventatively by spraying with tetramethrin (e.g. using the aerosol Garden Gun) for small plant collections (hold the can at least 30 cm from the plant when spraying), or with mercaptothion (e.g. Malasol) for larger collections.

All insecticides should, of course, be handled with care.

Opposite above: Caterpillars are best controlled with a carbaryl-based contact pesticide

Opposite below left: Gall midge fly damage on developing *Disa* flower buds

Opposite below right: The sticky surface of a Bug Trap is useful in controlling aphids

Diseases

Bacterial soft rot (*Erwinia carotovora*) is prevalent on the rootstocks, stems and foliage of evergreen and deciduous *Disa* species, especially during winter. It is recognized by soft, amber-coloured watery spots that turn brown and spread rapidly and can cause death of the plant within just a few days if not treated. Infection is usually the result of wounds, so care should be taken not to bruise plants. Affected plants should be isolated immediately symptoms are noticed, and the diseased parts cut away with a sharp blade, ensuring that the blade is disinfected after each cut by dipping it into a 2% formalin and 2% household bleach solution (sodium hypochlorite) for about 10 seconds. Alternatively dust the cut surfaces with flowers of sulphur, and place the plants in a cool dry spot to recover. Following correct watering procedure, and providing excellent drainage of the growing medium go a long way towards preventing bacterial disease, and it is advisable that the leaves of the deciduous species are kept as dry as possible, and that a well ventilated environment is maintained at all times.

Fungal diseases like *Ascochyta, Botrytis* and *Fusarium* attack the tuberoids, roots and leaves of both evergreen and deciduous disas. Infection in deciduous disas and other deciduous terrestrials like *Bartholina* and *Holothrix* is often the result of watering during the plant's dormant period, therefore strict adherence to relevant growth/dormancy cycles will reduce the incidence of fungal disease. In addition, the leaves of deciduous species should be kept as dry as possible. In evergreen species, cleanliness in the growing area and strict maintenance of good hygiene such as removal of dying leaves from the base of the plants, avoidance of over-watering and preventative spraying with suitable fungicides will greatly reduce the incidence of fungal disease.

In evergreen disas, an infected plant can be lifted, treated with a fungicide and re-planted in a sterile medium, but in the case of deciduous disas, treatment of fungal disease is much more difficult as fungicides may kill the mycorrhizal fungus upon which the plant is dependent, and therefore prevent exchange of nutrients to the orchid. If, in the latter instance, it is the tuberoid that has been affected, it is best to cut away the infected parts, dust the cut surfaces with flowers of sulphur, re-plant into dry soil containing the mycorrhizal fungus and allow the plant to go dormant. For evergreen disas, systemic fungicides containing benomyl (e.g. Benlate) should be avoided, but contact fungicides like those based on mancozeb (e.g. Dithane M45) and captab (e.g. Kaptan) can be used at half the recommended strength in preventative spraying to control *Asochyta* which causes black lesions on *Disa* leaves, while dusting with flowers of sulphur is recommended for control of *Botrytis* and *Fusarium* infection on rootstocks and roots. Plants undergoing treatment should always be isolated from other healthy plants until they have fully recovered, and infected parts should be burnt.

Viral diseases are spread mainly by insects that transfer infected plant sap from one plant to another, and result in deformed leaves and flowers. No remedy exists for viral disease in plants and those suspected of being infected must be isolated as soon as possible. If the characteristic symptoms of streaking or mottling of the leaves persist, the plants must be destroyed, preferably by burning. Viral diseases are spread mainly by insect vectors like aphids and mealy bugs, and also by slugs and snails, so control of these pests is of utmost importance. Infected sap can also be spread by humans from one plant to another on cutting instruments like secateurs, and it is of utmost importance that instruments are sterilised when used on one plant and then another. This can be done by placing instruments in a solution of 2% formalin and 2% hypochlorite (household bleach) for about 10 seconds, before being used on another plant.

Opposite above: *Disa cardinalis* flowering alongside a stream in the Langeberg Mountains in the southern Cape (see also page 50)

Opposite below: *Disa tenuifolia* (see page 75)

Disa bivalvata is a deciduous, winter-growing species

REFERENCES AND FURTHER READING

Bolus, H. 1888. The orchids of the Cape Peninsula. *Transactions of the South African Philosophical Society* 5(1). Darter, Cape Town.

Burnett, H.C. 1974. *Orchid diseases*. Department of Agriculture and Consumer Services, Florida.

Bytebier, B. 2004. *Disa barbata. Plant Talk* 35: 32-33.

Clements, M.A. and Ellyard, R.K. 1979. The symbiotic germination of Australian terrestrial orchids. *American Orchid Society Bulletin* 48(8): 810-816.

Clements, M.A. 1982. Developments in the symbiotic germination of Australian terrestrial orchids. *Proceedings of the 10th World Orchid Conference*. South African Orchid Council, Johannesburg.

Clements, M.A., Cribb, P.J. and Muir, H. 1986. A preliminary report on the symbiotic germination of European terrestrial orchids. *Kew Bulletin* 41(2): 437-445.

Clements, M.A. 1988. Orchid mycorrhizal associations. *Lindleyana* 3: 73-86.

Crous, H.T. 1997. Cultivating the coveted yellow *Disa uniflora. South African Orchid Journal* 28: 70-72.

Crous, H.T. 1998. Yuppie *Disa. Veld & Flora* 84(4): 134.

Crous, H.T. 2003. Some habitat observations on *Disa cardinalis. Orchids South Africa* 34: 24.

Cywes, S., Cywes, M. and Vogelpoel, L. 1987. *Disa* Veitchii 'Golden Petals' and 'Ted Schelpe' - clue to the genetic transmission of yellow pigments. *Veld & Flora* 73: 53-56.

Dressler, R.L. 1981. *The Orchids – natural history and classification*. Harvard University Press, Cambridge, Massachusetts.

Dressler, R.L. 1993. *Phylogeny and classification of the orchid family*. Dioscorides Press, Portland, Oregon.

Duncan, G.D. 2000. *Eulophia horsfallii* at Kirstenbosch. *Veld & Flora* 86(1): 16-18.

Duncan, G.D. 2002. *Grow nerines* p. 21. Kirstenbosch Gardening Series. National Botanical Institute, Cape Town.

Duncan, G.D. 2005. *Bartholina burmanniana. Curtis's Botanical Magazine* 22(1): 25-31.

Du Plessis, N.M. and Duncan, G.D. 1989. The family Orchidaceae. *Bulbous plants of southern Africa* p. 157-174. Tafelberg, Cape Town.

Dyer, R.A. 1954. *Disa uniflora. The Flowering Plants of Africa* 30, t.1180.

Dyer, R.A. 1954. *Herschelia graminifolia. The Flowering Plants of Africa* 30, t.1172.

Hall, A.V. 1973. *Herschelia charpentierana. The Flowering Plants of Africa* 42, t.1673.

Hall, A.V. 1973. *Herschelia hians. The Flowering Plants of Africa* 42, t.1674.

Hitchcock, A.N. 1989. Cultivation of *Disa uniflora*. Kirstenbosch National Botanical Garden pamphlet.

Johnson, K.C. 1969. *Disa uniflora* and its hybrids. *American Orchid Society Bulletin* 38: 135-146.

Johnson, S.D. and Linder, H.P. 1995. The systematics and evolution of the *Disa draconis* complex (Orchidaceae).

Botanical Journal of the Linnean Society 118: 289-307.

Johnson, S.D., Steiner, K.E. and Kaiser, R. 2005. Deceptive pollination in two subspecies of *Disa spathulata* (Orchidaceae) differing in morphology and floral fragrance. *Plant Systematics and Evolution* 255: 87-98.

Keeley, J.E. 1993. Smoke-induced flowering in the fire lily *Cyrtanthus ventricosus*. *South African Journal of Botany* 59 (6): 638.

Kurzweil, H. and Johnson, S. 1993. Auto-pollination in *Monadenia bracteata*. *South African Orchid Journal* 24: 21-22.

Kurzweil, H. 1994. The unusual seeds of the *Disa uniflora*-group, with notes on their dispersal. In: Pridgeon, A. *ed.* *Proceedings of the 14th World Orchid Conference, Glasgow*, pp. 397-399. HMSO Publications.

Lighton, C. 1973. Cape Floral Kingdom. Juta, Cape Town.

Linder, H.P. 1980. *Disa cardinalis* Linder (Orchidaceae), a new species from the Cape Province. *Journal of South African Botany* 46: 213-215.

Linder, H.P. 1981. A revision of *Disa* Berg. excluding section *Micranthae* Lindl. *Contributions of the Bolus Herbarium* 9: 1-370.

Linder, H.P. 1981. *Disa cardinalis*. The *Flowering Plants of Africa* 46 t.1826.

Linder, H.P. 1981. *Disa tripetaloides* (L.f.) N.E.Br. subsp. *tripetaloides*. *Veld & Flora* 67(3): 79-80.

Linder, H.P. 1988. *Herschelianthe lugens* var. *nigrescens*. The *Flowering Plants of Africa* 50, t.1977.

Linder, H.P. 1990. Hybrids in *Disa* (Diseae– Orchidoideae). *Lindleyana* 5: 224-230.

Linder, H.P. and Kurzweil, H. 1999. *Orchids of southern Africa*. Balkema, Rotterdam.

Lindley, J. 1830-1840. The genera and species of orchidaceous plants. London.

Lindquist, B. 1960. The raising of *Disa uniflora* seedlings in Gothenburg. *American Orchid Society Bulletin* 34: 317-319.

Linnaeus, C. 1753. *Species plantarum*. Laurentius Salvius, Stockholm.

Linnaeus, C. (*fil.*) 1781. *Supplementarum plantarum*. Holmiae, Brunsvigae.

Marloth, R. 1913. *The Flora of South Africa*. Darter, Cape Town.

Pauw, A. and Johnson, S.D. 1999. *Table Mountain, a natural history*. Fernwood Press, Cape Town.

Pienaar, R. de V., Littlejohn, G.M., Hill, M.S., Du Plessis, C. and Cywes, S. 1989. Chromosome numbers of various *Disa* species and their interspecific and complex hybrids. *South African Journal of Botany* 55(4): 394-399.

Reid, J. 1998. Gardening for butterflies: attracting the Table Mountain beauty (*Aeropetes tulbaghia*) to your garden. *Veld & Flora* 84(4): 135.

Roitman, G.G. and Maza, I.M. 2002-2003. The ecology and cultivation of terrestrial orchids of Argentina. *Herbertia* 57: 49-55.

Schelpe, E.A. 1966. *An introduction to the South African orchids*. Purnell, Cape Town.

Above: *Disa* Kirstenbosch Pride (see page 55)

Stewart, J and Hennessy, E.F. 1981. *Orchids of Africa: a select review.* Macmillan, Johannesburg.

Stewart, J., Linder, H.P., Schelpe, E.A. & Hall, A.V. 1982. *Wild orchids of southern Africa.* Macmillan, Johannesburg.

Stoutamire, W. 1977. *Disa uniflora* and *Disa* Veitchii. *American Orchid Society Bulletin* 46: 438.

Stoutamire, W. 1990. *Disa* cultivation in Ohio. *American Orchid Society Bulletin* 59(8): 803-809.

Stoutamire, W. 1990. Disas in South Africa. *American Orchid Society Bulletin* 59(8): 782-785).

Thompson, D.I., Edwards, T.J. and Van Staden, J. 2002. Conquering coat-imposed seed dormancy in the South African *Disa* (Orchidaceae): old ideas and new techniques. *South African Journal of Botany* 68(2): 242-281.

Thompson, D.I. 2003. Conservation of select South African *Disa* Berg. species (Orchidaceae) through *in vitro* seed germination. PhD thesis, University of Natal, Pietermaritzburg.

Thunberg, C.P. 1794. *Prodromus plantarum capensium.* Uppsala.

Vogelpoel, L. 1980. *Disa uniflora*: its propagation and cultivation. *American Orchid Society Bulletin* 49: 961-974.

Vogelpoel, L. 1980. *Disa* species and their hybrids. *American Orchid Society Bulletin* 49: 1084-1092.

Vogelpoel, L. 1984. *Disa uniflora - its propagation and cultivation.* Disa Orchid Society of South Africa, Cape Town.

Vogelpoel, L. 1986. Flower colour: an appreciation. *South African Orchid Journal* 17(3): 109-114.

Vogelpoel, L. 1887. New horizons in *Disa* breeding. The parent species and their culture 1. *Orchid Review* 95: 176-181.

Vopgeloel, L. 1993. The blue disas: Part 1. *South African Orchid Journal* 24(3): 66-72.

Vogelpoel, L. 1993. The blue disas: Part 2. *South African Orchid Journal* 24(4): 98-104.

Vogelpoel, L. 1993. A cultural calendar for disas based on seasonal biorhythms. *South African Orchid Journal* 24: 37-40.

Vogelpoel, L. 1994. Growing herschelianthes. *South African Orchid Journal* 25(2): 62-65.

Vopgelpoel, L. 1995. Some observations on pigments of plastid origin. *South African Orchid Journal* 26(2): 55-59.

Vogelpoel, L. and Cywes, S. 1998. The role of anthocyanin albinism in yellow *Disa uniflora* and other yellow flowers. *South African Orchid Journal* 29(2): 36-44.

Vogelpoel, L., Van der Merwe, D.W. and Anderson, B.D. 1985. A white form of *Disa racemosa*: a rare mutant with a great future. *American Orchid Society Bulletin* 54: 47-51.

Warner, B. and Rourke, J. 1996. *Flora Herscheliana.* The Brenhurst Press, Houghton.

Winter, J.H.S. 1981. *Disa uniflora* at Kirstenbosch. *Veld & Flora* 87(3): 74–76.

Wodrich, K.H. 1997. *Growing South African indigenous orchids.* Balkema, Rotterdam.

Opposite: *Disa uniflora* artificial hybrid raised at Kirstenbosch

GLOSSARY

aerobic living or functioning only in the presence of air or free oxygen

anaerobic living or functioning in the absence of air or free oxygen

deciduous shedding foliage at the end of a growth season

dorsal pertaining to the upper surface, as in the uppermost sepal

emasculation removal of the anther from a flower in order to prevent self-pollination

endemic having a natural distribution confined to a particular geographical area

epiphyte a non-parasitic plant that grows well above the ground on another plant

evergreen maintaining green foliage throughout the year

fynbos a derivative of the Dutch 'fijnbosch', referring to the community of small shrubs, evergreen and herbaceous plants and bulbs confined to the south-western and southern Cape

geophyte a perennial, usually deciduous plant with regenerative buds attached to subterranean storage organs, as in certain *Disa* species

hydrochory dispersal of seeds by water

hysteranthous leaves appearing after the flowers

inflorescence the arrangement of the flowers

inorganic not derived from living organisms, e.g. from chemical compounds

in vitro taking place outside a living organism, i.e. in a culture vessel like a glass bottle

leach the process whereby soluble components are dissolved out of the soil by water and then washed away

lithophyte a plant that grows on stones

microclimate the climate prevailing in a small, defined area, which may differ substantially from the general climate of a particular area

monocotyledon a flowering plant producing a single cotyledon or seed leaf

morphology the study of form, particularly external structure

mutualism a form of symbiosis in which two organisms exist in a close relationship of mutual benefit

mycelium thread-like, microscopic growth produced by fungi

mycorrhiza a symbiotic union in which the mycelium of a fungus penetrates the root/tuberoid in orchids and provides it with nutrients

organic relating to or derived from living organisms

palynology the study of living and fossil pollen and spores

pedicel the stalk of an individual flower

peduncle the stalk of an inflorescence

pollinium a cohesive mass of pollen grains

proteranthous flowers appearing immediately after leaves have died back

protocorm a minute, usually green body that develops from a germinating orchid seed

sepals the three outer floral segments in orchid flowers

stolon branch that bends towards the ground, rooting at the point where it touches the ground

symbiotic two organisms living together in which both obtain mutual benefit

synanthous flowers appearing together with the leaves

terrestrial growing on or under the ground

tuberoid the subterranean storage organ in *Disa* species and other terrestrial orchids

Table Mountain form of *Disa uniflora* (see page 44)

Index

Page numbers in **bold** denote illustrations

Disa cardinalis flowering in the Langeberg Mountains in the southern Cape

Hildegard Crous was awarded the Kirstenbosch Scholarship in 1990 and eighteen months later she joined the Kirstenbosch horticultural staff. Her duties included the care and maintenance of the *Disa* collection and it was here that her love of these plants was awakened. A tissue culture laboratory was established where she successfully propagated the rare yellow form of *Disa uniflora*.

She has presented several lectures on the cultivation and propagation of disas to local garden clubs, as well as to the British Orchid Society Conference in Wales, in 2003, and the European Orchid Council in Italy, in 2006. Hildegard spent three weeks tutoring students in practical tissue culture in the Ministry of Parks and Wildlife in Mauritius in 2000.

After leaving Kirstenbosch, she established her own private laboratory (registered with Cape Nature) at her home in Barrydale in the southern Cape. Current projects includes the propagation of the highly endangered *Disa barbata* as part of the '*Disa barbata* Project', headed by Benny Bytebier of the University of Stellenbosch.

Graham Duncan is a specialist horticulturist at Kirstenbosch Botanical Garden where he curates the collection of indigenous South African bulbs and deciduous disas, and the display inside the Kay Bergh Bulb House of the Botanical Society Conservatory.

His numerous popular and scientific articles on bulbs have appeared in leading horticultural and botanical journals, and he is the author of several titles in the Kirstenbosch Gardening Series. In 1989 he co-authored two major publications on indigenous bulbs, *Spring and Winter Flowering Bulbs of the Cape* with Barbara Jeppe, and *Bulbous Plants of Southern Africa* with Prof. Niel du Plessis, illustrated by Elise Bodley.

His special interest in the genus *Lachenalia* resulted in the publication of a popular guide to the genus in 1988 titled *The Lachenalia Handbook*, in the *Annals of Kirstenbosch Botanic Gardens* and an MSc (*cum laude*) in Botany from the University of KwaZulu-Natal in 2005. In 2001 he was honoured with the International Bulb Society's prestigious Herbert Medal.

Disa woodii flowering *en masse*, KwaZulu-Natal (see page 76)
(Photo: Gareth Chittenden)

USEFUL ADDRESSES

The Secretary
Disa Orchid Society of South Africa
(DOSA)
email: besgous@absamail.co.za
Tel: +27 21 913 6902

The Secretary
South African Orchid Council
P.O. Box 85
Edenvale 1610
South Africa
Tel/Fax: +27 11 452 0600
email: orchidcouncil@worldonline.co.za

The Secretary
Cape Orchid Society
P.O. Box 3347
Cape Town 8000
Tel: +27 21 855 5701

The Supervisor
The Seed Room
Kirstenbosch National Botanical Garden
Private Bag X7
Claremont 7735
South Africa
email: seedroom@sanbi.org
Tel: +27 21 799 8627
Fax: +27 21 797 6570

Right: *Disa bivalvata* in habitat, Table
Mountain (see also page 102)

Opposite: *Disa uniflora* in habitat,
Bainskloof

Wouter van Warmelo

For more information on South African plants visit

www.sanbi.org

click on

PlantZAfrica.com

Cameron McMaster

The hybrid *Disa* Kirstenbosch Pride 'Meg's Delight' (see page 50)

Opposite below: *Disa sagittalis*

Notes

Notes

Notes

Notes

Notes